Modern Electronic Security Systems

STEVEN HAHN

HAYDEN BOOK COMPANY, INC.
Rochelle Park, New Jersey

Library of Congress Cataloging in Publication Data

Hahn, Steven.
   Modern electronic security systems.

   Includes index.
   1. Electronic security systems. I. Title.
TH9737.H33   621.389'2   75-43662
ISBN 0-8104-5826-8

Copyright © 1976 by HAYDEN BOOK COMPANY, INC. All rights reserved. No part of this book may be reprinted, or reproduced, or utilized in any form or by any electronic, mechanical, or other means, now known or hereafter invented, including photocopying and recording, or in any information storage and retrieval system, without permission in writing from the Publisher.

Printed in the United States of America

    1 2 3 4 5 6 7 8 9   PRINTING
76 77 78 79 80 81 82 83 84   YEAR

G. R. Peirce
Professor of Elec. Eng'ng
Emeritus

Reviewed for "Choice"
Nov. 1976

Modern Electronic Security Systems

Preface

This book offers an overall view of electronic systems and techniques which are currently employed in the field of security. In doing so, a fair representation of manufacturers' products has been included to provide the latest available information in the business. Without the manufacturers' cooperation and ready assistance, this book would not have been possible, and I want to take this opportunity to acknowledge their help.

Special thanks are also extended to Mr. Leo Goodhart and Mr. Nat Verger of ADEMCO, Mr. Dennis Shapiro of Aerospace, Mr. Roy Stockdale of NAPCO, Mr. Paul Scheerer and Mr. Stuart Miller of Detectron Security Systems, and Mr. George Kuredjian, my patient editor.

Steven Hahn

East Hampton, N.Y.

Contents

Introduction *1*

1. Basic Security Systems *4*

    Intrusion Detectors • Controls • Local and Remote Alarm Indicators • Local Alarm Power Consumption • Remote Point Signaling

2. Perimeter Intrusion Detectors *27*

    Perimeter Switches • Magnetic Switches • Vibration Contacts • Step Mats • General Purpose Contacts and Plungers • Trip Wires and Traps • Foil • Tilt Switches • Panic Buttons

3. Space Intrusion Detectors and Signal Processing Circuits.. *35*

    Doppler Intrusion Detectors • FCC Regulations Regarding Space Intrusion Detectors • Antenna Radiation Patterns and Characteristics • Sonic and Ultrasonic Space Intrusion Detectors • Amplifiers • Stress-Sensitive Transducers • Seismic Transducers • Audio-Detection Intrusion Detectors • Photoelectric Intrusion Detectors • False Alarms • Signal Processing Circuits

4. Security Controls *65*

    Hard Wiring Security Circuits • Silicon Controlled Rectifiers (SCR) • Security Controls • Key Switches • Remote Control Stations • Electronic Time Delay Mechanism • Automatic Recycling Controls • Zone Selection Circuitry • Control Terminal Strips

5. Power Supplies and Standby Power 83

   Continuous Power Consumption • Transients and Noise • Ripple, Line Regulation, and Load Regulation • Protecting Against Overloads • Automatic Standby Battery Systems • Rechargeable Batteries • Battery Ampere-Hour Capacities

6. Local Alarms 99

   Parameters • Alarm Bells • Electronic Sirens • Mechanical Sirens • Other Local Alarms

7. Remote Alarms 113

   Tape Telephone Dialer • Digital Telephone Dialer • Telephone Subsets • Leased-Line System • Reversing Relay System • McCulloh Loop System • Remote Alarm Locations

8. Special Systems 137

   Integrated Systems • Wireless (Radio) Systems • Closed-Circuit Television

9. Installation of Security Systems 149

   Cables • Perimeter Devices • Decals • Special Tools

Index 167

Modern Electronic
Security Systems

Introduction

Not so very long ago, electronic security systems were of primary concern only to large institutions, such as banks or government agencies, where money or highly confidential data was kept. The only other type of application which electronic security systems have enjoyed in the past is their usage under special circumstances, as the protection of irreplaceable art objects or in a few retail establishments which deal in extremely expensive goods, such as diamonds, furs, stamps, etc. Thus, one can in all fairness say that the market for electronic security systems in the past has been primarily in those applications where the protection of very valuable goods was involved. In recent years, however, the market for security systems has dramatically changed as a result of large increases not only in burglary, but also in the kind of burglaries which take place. The usage of electronic security systems is no longer the sole concern of banks and jewelry stores, but now has become a realistic preoccupation for every businessman, large and small, as well as private citizens residing in apartments, modest homes, and elaborate homes. One need only read the local papers or national magazines to dramatically see the alarming rate at which crime has increased in recent years in the United States. Furthermore, notwithstanding wishful thinking, there appears to be no relief in sight. In addition, analysis of crime statistics will clearly show that the increase in burglaries has not dramatically changed with respect to banks or other high content value institutions. The most pronounced increase in burglaries has taken place in the home and small business level. In a like manner, the basic profile of the burglar has changed in the past few years. We are no longer primarily concerned with the skillful professional who plans his crime carefully using the most sophisticated techniques and employing skillful accomplices. Such burglaries at the "professional level" are, of course, still taking place, but the rate with which they are growing is much smaller than the rate at which "impulse intrusions" are taking place. It is the impulse type of intrusion which makes it mandatory for every individual to carefully consider and evaluate the usage of an electronic security system. The typical impulse intruder does not carefully plan his crime. Instead, we are dealing with an individual who for some reason—perhaps drug needs—suddenly requires money and simply on impulse selects the nearest house or business in order to obtain goods which can then be sold to the nearest fence. Such a burglar enters the premises, snatches a few key items, such as television sets, typewriters, photographic equipment, etc., then leaves as rapidly as he entered and the entire crime may take no longer than a few minutes. Furthermore, in many instances, stealing for the generation of cash does not even appear to be the motive. As we are all aware from reading the news and listening to the media, we are living in an era in which "motiveless crimes" are not uncommon. Thus, intruders will enter a home or a place of business not necessarily to steal, but simply to destroy, deface, damage, and otherwise ruin property. This type of crime may be combined with burglaries which are made for monetary gain. Someone will enter rapidly to steal a TV set, but will take a few minutes more to senselessly destroy and slash property, deface and ruin goods, and even set fire to the structure for no apparent reason. It is important to realize that the electronic security systems which are designed to prevent this type of "impulsive intrusion" must function quite differently from their more elaborate counterparts whose purpose it is to detect sophisticated burglars who may spend hours on the premises, systematically looting or opening vaults, etc. The modern security system which is designed to prevent impulse burglary must react very fast since the burglar is intent on maliciously destroying property and getting out as fast as possible. It will consequently no longer suffice for the system to automatically call the police since, even under the best of circumstances, the police cannot arrive fast enough to prevent the damage. In most cases, the response of the police, especially in these times when police are so overloaded with work, usually takes in excess of 10 minutes and by that time the impulse burglar has long gone, leaving behind him a trail of vicious damage. Thus, in the case of the impulse burglar, the user of electronic security systems can draw little comfort from the fact that the system automatically dials the police or some central service point in the event of an intrusion. No matter how efficient the law enforcement agencies are, it is most unlikely that they will be at the scene of the crime before the damage is done. As a matter of fact, in recent years, police resistance, at the local level, has become increasingly high to the usage of electronic security systems which dial the local police station. In view of the tremendous overload of work which policemen are currently carrying, one can well understand their objection to this type of security system. As a result, numerous towns and cities in the United States have passed ordinances which specifically prohibit the usage of security systems which, in the event of an intrusion, dial the local police. It consequently becomes the function of the modern security system to immediately frighten away the impulse burglar and at the same time alert neighbors, passersby, and patrolling policemen to the fact that an intrusion is taking place in the hope that an instant citizen action will take place, thereby flushing out the burglar. It is this type of security system which has enjoyed the greatest growth in recent years. Because we are here dealing with an unsophisticated burglar, often young and not necessarily an experienced criminal, these systems have proven to be quite effective. In the case of the sophisticated burglar who is after very large targets, even the very best security systems may be useless since no security system has ever been made which cannot be bypassed by a highly skilled criminal who has taken great pains and time to analyze the premises which he is about to violate.

In summary, this book attempts to describe in technical terms all the approaches which are currently available to industry and private individuals with respect to the installation and usage of electronic security systems. As is the case in all rapidly advancing, growing technologies, no single electronic security technique is an all-encompassing panacea. The user or installer of security systems must realize that each security application requires its own special approach with regard to the electronic equipment involved. Furthermore, the selection of such equipment cannot entirely be made on the basis of manufacturers' claims, which often are optimistic with respect to performance. It becomes necessary for users and installers of security systems to have a fairly thorough understanding of the technologies which are involved and of what they can and cannot deliver in terms of performance. The electronic security equipment field is fairly young and growing by leaps and bounds. As a result, a somewhat chaotic condition exists in which manufacturers are constantly appearing while others disappear. The net effect is that one may encounter equipment which was designed hastily, for which claims are made which have little basis in fact, and for which technological backup and application help is virtually nonexistent. Obviously, everyone would like to have a security system, especially an intrusion detection device, which would detect, on a 100 percent fail safe basis, any intruder or any unauthorized person. Furthermore, such a system should be able to differentiate between a friend or harmless passerby and an intruder. In addition, such a security system, ideally, must never give false alarms and must know the difference between humans and possibly false alarming signals, such as large dogs, wind rattling structural members, etc. Lastly, this ideal system ought to be reasonably priced. Like all ideals, such a security system simply does not now nor ever will exist and individuals concerned with electronic security must learn the capabilities of the technology, and the reputation of the manufacturers, so that they can exercise proper judgment and make sound product decisions. To do this it does become necessary to learn the technologies involved, and it is the purpose of this book to help towards that end.

1
Basic Security Systems

The basic function of a security system is to detect as fast as possible an unauthorized entry into a defined area. Furthermore, the ideal security system should also be difficult to bypass or override; must be highly reliable and operate under adverse conditions (for example, in the event of power failure); should not be subject to false alarming, and recognize the difference between an unauthorized human intruder and a harmless intruder, such as a dog, etc. Essentially, every security system can be divided into three generic equipment groupings (see Fig. 1-1): (1) the intrusion detectors; (2) the control panel; and (3) local and remote alarm systems.

Intrusion Detectors

The intrusion detector is the device which is used to initially detect an unauthorized entry. Such intrusion detectors can be subdivided into two basic categories: (1) perimeter detectors, and (2) space (volumetric) detectors.

Perimeter Intrusion Detectors

Perimeter intrusion detectors essentially consist of different types of switches which are strategically located around the perimeter of the protected premises. A classic example of a perimeter intrusion detector is the popular magnetic switch (Fig. 1-2) which can be mounted in any door or window. Such a device consists of two pieces—a magnet and a reed or slide switch

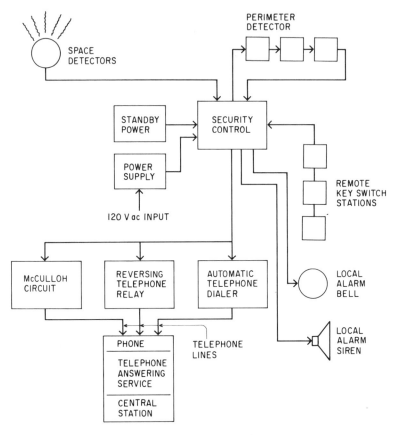

Fig. 1-1. Basic elements of a security system.

which is usually supplied in the electrically closed position. When the magnet is moved away from the switch as a result of the window or door being opened, the switch opens up triggering an alarm. Most perimeter detectors work on this basic switch principle and are employed in wide variety of applications.

Another type of perimeter switch is a mat or ribbon which is laid around or at the entrance point of the area to be protected. (Fig. 1-3).The mat is constructed in such a way that anyone stepping on it will close an electrical contact triggering an alarm. (A common use of this type of perimeter detector, though not for security purposes, is in a gas station where a car driving over such a device will trigger a bell to alert an attendant.)

Another common use of a perimeter detector is metallic window foil in which a conductive thin foil is glued to glass. When the glass is broken, the foil opens up the circuit triggering the alarm. A more sophisticated type of perimeter detector consists of a light beam employing a light source and a

BASIC SECURITY SYSTEMS 7

Fig. 1-2. Perimeter intrusion detectors utilizing magnetic switches.

Fig. 1-3. Mat intrusion detector.

photoelectric cell. Whenever the beam is broken by someone passing through it, an alarm circuit is fired. Still another type of perimeter detector is a vibration switch. This device has adjustable contacts which close when vibration is exceeded on a wall or a panel. Vibration perimeter switches are very useful in protecting roofs or walls where the intruder might employ a saw or a hammer to gain entry. Note that these perimeter detectors, though taking various forms, have one thing in common in that they protect the circumference of an area, i.e., they do not offer protection in three dimensional (volumetric) space. Furthermore, as most intruders know, a perimeter type of detector can easily be "jumped." Such jumping is accomplished merely by paralleling the electrical connection prior to breaking it. Thus, by putting a jumper wire across a normally closed magnetic switch prior to entry (this might be done by sawing a hole through the door) the effectiveness of a magnetic switch can completely be defeated by electrically shunting it out. Normally electrically open perimeter detectors (i.e., circuits are closed upon intrusion) are even more easily defeated by simply cutting the wires leading to them. Consequently, such devices offer less and less protection with regard to the prevention of robbery and unauthorized entry. Furthermore, since perimeter detectors must be interconnected into a wired system, their installation cost, especially if many such devices are used, tends to become very high. This is due to the fact that a wire has to be run to each perimeter transducer and sometimes such wiring involves a considerable amount of labor (for example, in a

Fig. 1-4. Doppler space intrusion detector.

home where the cable, for esthetic reasons, must be hidden). For these reasons, the popularity of perimeter detectors, such as magnetic switches, step mats, window foil, and trip wires, has diminished and their usage has to a degree been superseded by the employment of space intrusion detectors (see Fig. 1-4).

Space Intrusion Detectors

Space intrusion detectors are devices which can detect an intrusion in volumetrical (three dimensional) space. In operation, a space intrusion detector generates an energy field (usually radio-frequency or sonic) in all directions. When this energy field is disturbed, as by the entry of a human being or other mass, a signal is initiated which triggers the alarm. Obviously, a space intrusion detector is most difficult to jam because one has to pass through its energy field to get at the device.

Many space detectors employ the Doppler principle. This premise relates to the slight change in frequency of the signal emitted from a transmitter in motion, or if the distance traveled by a reflected signal should be suddenly changed. The frequency of the signal increases when the transmitter is moving towards the receiving site, and decreases if the transmitter is moving away from the receiver. Similarly, the frequency of a reflected signal would increase if a moving object entered the energy field. A simplified example of the Doppler principle is a train approaching a railroad station. The whistle sounded by the approaching train will have a higher pitch or frequency than when the train is stationary in the station. For the same reason, the whistle sounded by a departing train will have a lower pitch.

Another example of the Doppler effect is the police radar speed detector. Its transmitter emits a signal in the 1000-megahertz range (1 gigahertz or 1GHz). An automobile entering this energy field at 60 mph will reflect the signal back to the radar speed detector with a change in frequency of approximately 600 hertz (Hz) with respect to the original frequency transmitted at 1GHz. This frequency change is called the Doppler shift. In the security field, a whole series of detectors have been developed which operate in various segments of the acoustic and electromagnetic spectrum utilizing the Doppler principle. They include:

1. Sonic devices operating between 4,000 and 8,000Hz.
2. Ultrasonic devices operating between 15,000 and 40,000Hz.
3. Ultra-high frequency devices operating at 915 million hertz (915MHz) which requires FCC certification.
4. Microwave devices operating at 2.5 and 10 billion hertz (1.5 and 10GHz) which also require FCC certification.

A detailed explanation of the Doppler principle is presented in Chapter 3.

Each of the Doppler type intrusion detectors have unique characteristics which make them ideally suited for specific applications. (Table 1-1 shows in summary fashion, the overall characteristics of each type of space detector.) For example, detectors using sonic energy actually can be heard within the environment in which they are used. Sound is fairly nondirectional and does not pierce opaque objects, such as walls, windows, etc. Furthermore, sound losses in air are not very high so that a sonic Doppler detection device can cover moderately sized areas up to approximately 20 ft x 20 ft.

The ultrasonic detector operates almost in the same manner as a sonic detector, except that its energy is in the ultrasonic region (approximately 20kHz-45kHz) and cannot be heard by the human ear. Ultrasonic energy will not pierce structural members, such as walls. Moreover, the losses of ultrasonic energy through air are quite high. Consequently, ultrasonic devices generally do not cover as large an area as sonic devices, though usually not affected by audible sounds, can be falsely triggered by hot air currents and some sonic sounds which have ultrasonic components, such as the hiss of a radiator, jet engine noise, or certain telephone bells.

Space detection devices operating in the Doppler mode also are available utilizing ultra-high (915MHz) or microwave (2.5GHz and 10GHz) frequencies. Here the energy may be omnidirectional, or in the case of 10GHz devices, can also be beamed in a fairly narrow fan-shape pattern. The energy in this type of detector does tend to pierce walls and glass (especially at the lower frequencies, 915MHz). Generally speaking, only metallic surfaces, such as copper and aluminum screening, foil insulated walls, and wire lath, will stop this type of energy. In addition, large metallic objects in the protected premises, such as machinery, refrigerators, ducting and metal cabinets, will deform this type of energy field and will change the coverage pattern. Signals of this type, unlike sonic and ultrasonic energy, are not affected by sounds or by hot air currents. However, they may, on occasion, be affected by strong nearby radio signals, such as are emitted by police cruisers, nearby transmitters or airport communication equipment, etc.

Not all space intrusion detectors employ the Doppler technique. One of the most unusual non-Doppler devices employs a microphone in the premises which are to be protected. The signal from this microphone is sent back, usually via telephone lines, to a central monitoring point where electronic circuitry is used to listen to the sounds in a protected premises and initiate an alarm when suspicious noises occur. Such devices often have the capability, at the remote listening point, of opening up the microphone circuit so that a guard can actually listen to all the sounds that are emanating from the protected premises and discern whether or not an intrusion is taking place. Space intrusion detectors are also available which operate on the capacitive principle where electronic circuitry will detect the presence of another mass within a moderately small radius of 5 or 6 feet. In such a system, the intruder acts

Table 1–1. Space Intrusion Detector Characteristics

Device	Operational Mode	Pattern	Range	Comments
Sonic (Audio Detection)	Listens to sound	Omni-directional	25-ft radius	Does not go through walls, partitions, etc.; can sometimes be triggered by noises from outside traffic, telephone ringing, etc.
Sonic	4,000–8,000 hertz transmitted and reflected Doppler signal	Omni-directional	25-ft radius	Does not go through walls, partitions, etc.; may be triggered by moving objects (chandeliers, venetian blinds).
Ultrasonic	15,000–40,000 hertz transmitted and reflected Doppler signal	Omni-directional	15-30-ft radius	Does not go through walls, partitions, etc.; may be triggered by moving objects chandeliers, venetian blinds).
Ultra-high frequency (RF)	915MHz transmitted and reflected Doppler signal	Omni-directional	20-ft radius	Will go through most walls, partitions, etc. blocked by metal; may trigger on moving objects within the protected area (chandeliers, venetian blinds); may also trigger on external radio signals emanating from police transmitters.
Microwave (RF)	2.5GHz-10GHz transmitted and reflected Doppler signal	Fan-shaped team	Up to 75 ft	Goes through walls, partitions (less so at 10 GHz), can trigger on moving objects within the protected area (chandeliers, venetian blinds, and interfering radar systems at airports and military installations); may be used outdoors.
Infrared	Passive detector senses heat rise	Omni-directional or wide fan shape	15-ft radius	Will not go through walls, partitions, etc.; may trigger on hot objects (radiators, incandescent bulbs, rising sun).
Stress	Resistive changes resulting from flexure in structural members	Omni-directional	15-ft radius	Transducer epoxied to structural member which flexes; not affected by extraneous signals, heat, noise, etc.
Capacity	Change in capacity in a tuned circuit	Omni-directional	3-ft radius	Not affected by extraneous signals, heat, noise, etc.; excellent for short range detection.
Ambient light	Light level changes seen by photoelectric cell	Omni-directional	10-ft radius	Not affected by extraneous signals, heat, noise, etc.; excellent for short range detection; needs lighted premises.
Closed circuit TV	TV camera system	What camera sees	Up to 50 ft from camera lens	Not affected by extraneous signals, heat, noise, etc.; needs lighted premises.
Seismic	Low frequency vibration in ground	Omni-directional	50-ft radius	May be triggered by ground tremors caused by trucks, thunder, noise; excellent for outdoor protection.

12   ELECTRONIC SECURITY SYSTEMS

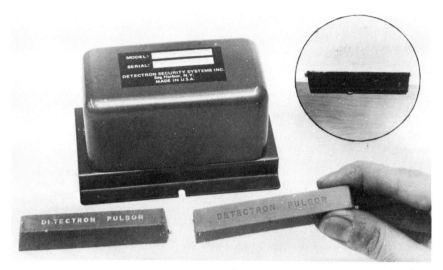

Fig. 1-5. Pulsor stress detector intrusion units. (Insert shows detector cemented in place.)

as the second plate of a capacitor and the capacitive change which takes place in the detector circuit triggers an alarm. This type of intrusion detector is ideally suited for protecting small safes and sensitive filing cabinets.

Another effective space intrusion detector is a system which employs a minute crystal (Fig. 1-5) which is cemented to structural members such as beams, risers, stairwells or any structure which flexes when an unwanted intruder walks in the premises. These stress detectors are made in such a way that an electrical resistance change takes place as a result of structural member flexure. This change results in a signal which is then fed back to special amplifiers which trigger the alarm. Such stress intrusion detectors will detect people walking over comparatively large areas having as much as 40 feet in diameter. A number of stress sensitive detectors can be combined to protect very large areas with very low false alarm rates because they will not falsely trigger on air currents, sounds, outside radio fields, etc. In addition, large size areas can be protected providing one can get underneath the space in order to epoxy the stress transducer to the structural member.

A number of other unusual intrusion space detection devices are available which, like the Pulsor stress detectors, do not operate on the Doppler principle. One of these is the infrared detector which "sees" the heat emitted by the human body as it enters a room, thereby triggering an alarm. Such detectors generally work over short ranges only—up to a 15-foot radius. Their fields are, of course, invisible and they are not affected by sound or radio interference. In their electronic circuitry they employ special techniques which must be used in order to differentiate between the infrared energy (heat) emanating from a human body and the heat generated by such false alarm causing signals as an incandescent light bulb, radiators going on and the sun rising as seen by the infrared detector through a window.

Another non-Doppler device is the ambient light intrusion detector which looks at the fixed light level in any environment. When shadows are cast, as a result of movement, in the protected area, the alarm is triggered. This device, though working on photoelectric principles, is quite different from the perimeter type photoelectric light beam which must be broken in order to trigger an alarm. Ambient light detectors need some form of illumination to operate and will not work in total darkness. Furthermore, the distances from which they can detect moving shadows are comparatively small, usually up to 10 feet.

Still another method of space detection employs a TV camera and a remote TV screen. Special photoelectric detectors are mounted to the remote TV screen and these detectors will sense motion in the protected area which the camera is viewing. Here again, just as was the case with the ambient light detector, some form of illumination is required. However, the area which can be covered by such a device could be quite large, and is only limited by the field of view of the TV camera.

In most space intrusion detectors, whether Doppler or not, some form of energy is radiated by the detector. The change in energy, as the result of a moving mass within the energy field, is sensed by the detector resulting in an alarm condition. Some detectors, infrared and stress, do not actually transmit energy but look for the energy generated by the intruder. In the preceding paragraphs, we have given some indication of the range of each generic space intrusion detection method. In many instances, the size of the protected area can be increased markedly through the use of numerous intrusion "transceivers" all operating through a single electronic amplifying system. For example, a single ultrasonic transmitter-receiver may only be able to cover a 10 ft x 20 ft area. However, as many as twenty such ultrasonic transceivers may be used together, all operating through a single amplifier, with the result that very large areas can be covered.

One of the major problems with any security system, and especially with space intrusion detectors, is the matter of false alarms. False alarm signals can be generated in intrusion detectors by a wide variety of causes depending upon the operational mode of the detector. In order to limit and eliminate false alarming, space intrusion detectors employ sophisticated electronic circuitry (as discussed in Chapter 3) to process the basic intrusion signals received from the detector prior to giving an alarm signal. This most important signal processing activity can take a number of forms. For example, in Doppler devices, the actual frequency of the Doppler signal is dependent upon the basic transmitter frequency and the velocity with which the intruding mass moves. The transmitting frequency is, of course, fixed and so the only variable is the velocity of the intruder. If we take the case of a space intrusion detector operating at 915MHz, the Doppler signal generated by a man walking at a normal rate through such a field is approximately 1 to 10 hertz above or below the basic transmitter frequency depending on whether or not the man is moving towards the transmitter or away from the transmitter. A bird or a car entering such a field would generate a much higher Doppler frequency since the velocity of these masses is considerably more than the velocity of a walking intruder. This fact is of great importance in limiting false alarms in space intrusion detectors. If the Doppler frequency range is known, the bandpass characteristics of the associated amplifier can be tailored to reject frequencies which do not fall within the wanted range. This signal processing concept holds true even for non-Doppler devices such as the stress detectors. In this type of device, we are dealing with resistive changes produced by the weight of footsteps which slightly deform structural members. If the average walking rate of a human being is known, an amplifier can be designed whose frequency response is held closely to this rate. Another very effective method of signal processing can be applied to Doppler devices. As previously stated, the Doppler frequency is above or below the transmitted frequency depending on whether the object is moving towards or away from the transmitter. A false alarm-causing signal might be generated by a waving tree, a moving venetian blind or chandelier. These objects, since they move back and forth from a fixed position only, generate Doppler signals above and below the transmitted frequency. Consequently, if the Doppler frequency were sampled over a period of time, let us say a second or two, the algebraic sum of the shift would be nearly zero. Circuits can be designed which look for the sum of the Doppler frequency, rejecting—for alarm purposes—any signal which fails to have a pronounced positive or negative shift. Furthermore, many amplifiers for space intrusion detectors employ integration and event counting circuitry which also helps to reduce the possibility of false alarms. In such circuits the intrusion signal is stored and summed over time and a number of intrusion events must take place within a given time period before the actual alarm signal is generated. The circuits can also be adjusted so that the amplitude of any single event is always held constant and each event gets only "one count" in the event-counting circuit, regardless of its initial amplitude. This type of circuitry becomes very useful in limiting false alarms by such troublesome signals as lightning. A nearby lightning stroke can easily induce a large pulse in the electronics of a sophisticated space intrusion detector. If the circuit contains integration and event-counting features, the lightning stroke would not trigger the alarm, since it is a single event which, in all likelihood, will not repeat in the time sampling period. Still another method of signal processing which is employed in sonic and audio detection devices uses a series of frequency selective filters which actually listen, through the microphone in the protected premises, to those frequencies (within the range of about 300 to 5,000 Hz) which are common and

BASIC SECURITY SYSTEMS   15

pronounced in human activity—for example, breathing and speech, etc. These circuits would reject frequencies which were of no interest, such as bells, outside traffic noise, etc.

Advances in modern electronic technology continue to generate more and more sophisticated space intrusion detectors and associated signal processing circuitry. This is an active, growing field and it is difficult to cover all the circuit innovations and techniques currently within the grasp of the technology. (For a summary of some of the general characteristics of the various types of space intrusion detectors, refer back to Table 1-1.) It should be pointed out however, that although space intrusion detectors are very difficult to override, they do have—as a result of their more sophisticated circuitry—a somewhat higher false alarm rate than their perimeter counterparts.

Controls

Once the alarm has been triggered by a detector, a signal (circuit opening or closing) indicating an alarm condition is sent back to the control. It is one of the functions of the control to take the alarm signal and process it so

Fig. 1-6. Alarm control unit for perimeter loop circuit.

that a series of events take place which result in the ringing of a bell, the sounding of a siren, and transmission of the alarm condition to either the police or some other control point. In addition, the control must also develop all the necessary voltages which are required by the entire system. Furthermore, these voltages are generated in the control in such a manner that standby power is automatically switched into the system in the event of ac power failure. Also, controls should offer important additional operating characteristics which make for a more reliable, flexible security system.

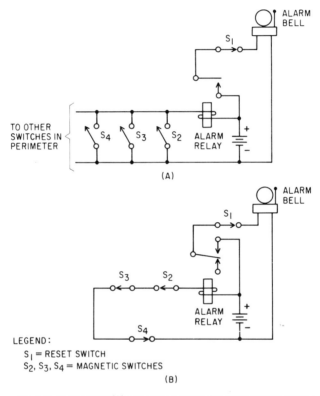

Fig. 1-7. Open-circuit (A) and closed-circuit (B) alarm control systems.

A basic alarm control circuit usually consists of a battery and relay arrangement as shown in Figs. 1-6 and 1-7. When the relay is energized by any of the intrusion detectors (magnetic switches S-2, S-3 or S-4) in either the open-or-closed-circuit arrangement, the alarm bell powered by the battery will ring. This type of alarm control, however, has drawbacks because it depends upon battery power and the life of a battery is limited. The use of an ac power supply, as indicated in Fig. 1-8, will overcome this deficiency.

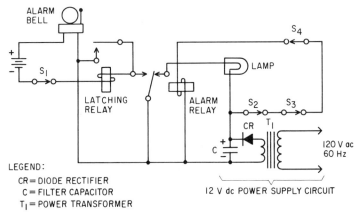
Fig. 1-8. Latching relay alarm control circuit.

Another drawback is that alarm control circuits only sound the alarm as long as any intrusion detector is triggered. Consequently, quickly opening and closing a door or window, protected by a magnetic switch or similar device, would cause the alarm to only ring momentarily, thereby making it virtually useless. This condition can be eliminated by utilizing a latching relay control circuit as shown in Fig. 1-8. In this type of circuit, the control relay locks up or latches whenever operated, even momentarily, by the triggering of an intrusion detector. The alarm bell, therefore, will ring continuously until reset by the manual opening of switch S-1. One very useful variation of this circuit will ring the alarm bell for a fixed amount of time (perhaps 5 or 10 minutes) and then automatically rearm itself, providing the integrity of the intrusion detection circuit has been restored. This feature is very useful in instances where the alarm might be triggered falsely by wind gusts momentarily rattling a door which is protected by a magnetic switch, causing the switch to open for a fraction of a second, thereby setting off the alarm. If the control has the automatic recycling feature, it would not be necessary for someone to return and reset the control, since this function is accomplished automatically after the alarm period has elapsed.

In all instances, whether the control permanently latches in the alarm state or recycles automatically, a key switch station must be provided from which authorized personnel can energize and reset the entire security system. The key switch station, which may be located on the control panel or at a remote point, and must use a good mechanical-electrical lock (Fig. 1-9) so that only authorized personnel with keys can control the system. Furthermore, these control points must have some indicating method to show the status of the system. Such indicators are usually in the form of pilot lights or meters which show the actual condition of the system, whether the perimeter loop is open or not, status of the standby power supply, alarm conditions, trouble

Fig. 1-9. Examples of remote control and panic-button stations.

conditions, etc. Many security applications require that the system be armed or reset from a number of remote positions, so the control must have multiple key switch station capabilities. This feature is especially important in plants and businesses where a number of exits and entrances are used by authorized personnel. Multiple key switch operation is also important in the home where the system owner energizes the system not only when leaving, but also upon retiring, so that the premises are protected while the owner is asleep.

It is also the function of the alarm control equipment to furnish necessary power to all the intrusion detectors as well as to the bells and sirens, dialers, etc. In some larger security installations, the amount of power required for this type of service can be considerable, running into the 5-ampere region. Consequently, a good security control unit should have ample power supply reserves to run the entire system, both on a continuous basis where comparatively little current is drawn, and on an alarm basis where heavy currents are drawn. In addition, it is most advisable that the control equipment have automatic switchover to standby power capabilities. Standby power can be in the form of dry cells or continuously charged, rechargeable standby batteries. Furthermore, it is desirable that the control unit contains at least a number of isolated sections of relay switching so that a wide variety of alarm devices can be independently handled, including bells, sirens or automatic dialers. In addition, a control unit should be easily mountable and come in a sturdy case with internal electrical tamper proofing so that in the event the intruder can get to the control equipment, it would be impossible to destroy it before the alarm has sounded.

BASIC SECURITY SYSTEMS   19

Local and Remote Alarm Indicators

Once an intrusion signal has been processed by the control, alarms must be actuated which indicate that an intrusion is in progress and these devices can be divided into two basic categories: (1) the local alarm system and (2) the remote alarm system.

Local Alarm System

The local alarm system must immediately warn the intruder that his presence has been detected and also attract the attention of others (police, guards) that an intrusion has taken place. Local alarms take the form of bells and sirens and such alarms should obviously be as loud as possible and have the highest probability of attracting attention. Furthermore, they should be so loud as to make it psychologically uncomfortable for the intruder to continue his destructive efforts.

Remote Alarm System

The remote key switch stations are often located at major entrances and exits where they might well be seen by a prospective intruder. Even though these remote key switch stations should be internally protected against tampering, a sophisticated intruder might pick the lock, thereby shutting off the entire security system. Modern controls circumvent this disadvantage by locating all remote key switch stations within the protected area so that in order to turn the system on or off, one has to actually violate one of the intrusion detectors (space or perimeter). Since, with this system, authorized personnel energizing the system will pass through an intrusion detector after they have armed the system, it becomes necessary to have circuitry which will not trigger the alarm when authorized personnel leave or enter. This is accomplished by adjustable time delay circuits which permit a certain amount of time (usually adjustable from 0 to 60 seconds) to elapse before the alarm sequence commences after one of the intrusion detectors has been violated. Thus, authorized personnel entering in the morning might violate a space intrusion detector located in the lobby of the plant. They then would have, for example, 30 seconds to get to the key switch station to turn the system off before an alarm will result. Even if a thief knew of the 30-second time delay of such a system, it would be highly unlikely that he could get to the remote key station and pick the lock in the 30 seconds which had been allocated for this function. In a reverse manner, authorized personnel leaving the premises at night can arm the control at the key switch station and then have 30 seconds to get out of the plant during which time they can actually violate intrusion detectors without triggering the alarm.

In addition to delay circuitry, a number of controls have two intrusion alarm outputs. The first one sounds immediately the moment that any detector is violated, while the second commences after a time delay has expired.

With such a technique, the first alarm is usually in the form of a low level local buzzer warning authorized personnel that the alarm system has been activated and also advising them that the alarm system is operational, thereby giving a self-checking feature when leaving or entering. Once the first alarm has sounded, authorized personnel has a fixed time period to shut off the security system before the second alarm, the bell and siren, will go off.

Another very desirable feature found in a number of controls is the so-called panic button. This is a special form of remote station which contains a single button located at various strategic places throughout the premises. When this button is pressed, regardless of the setting of the system control station key switch, an immediate alarm condition will result. Such a button has very important usage during holdups in the daytime when the system is not on or even at times when the system is on and an intruder enters in some way without triggering the protective detectors. These buttons are usually located in such areas as teller windows in banks, cashier cages in stores, bedrooms in homes and other strategic areas where instant energizing of the security system may prove desirable.

Loudness of Alarms: Decibels

The loudness of alarm devices normally is rated in decibels (dB) as measured at a given distance in free air from the center axis of the particular device. A typical electronic siren, for instance, may have a loudness rating of 110 dB at a distance of 10 ft. A 10-inch fire bell usually is rated at 102 dB at the 10 ft distance. The term decibel has been used for many years in telephone communications and in the electronics field to express logarithmically the ratio between two values of voltage, current, power or sound levels. The decibel, as a logarithmic unit, closely indicates the response of the human ear to sound waves. The ear responds approximately proportional to the logarithm of the energy of the sound wave and not to the energy itself.

It should be pointed out that the dB loudness value must be measured by suitable instruments, in accordance with standard procedures, to be meaningful. Different bells and sirens have diverse sound radiating patterns which can result in unequal loudness ratings at various audio frequencies. Consequently, the dB loudness figures cited by manufacturers should not be the sole data considered in selecting an alarm device.

The sound-pressure level of an alarm device, such as a fire gong, can be expressed in decibels referenced to a standard value. In this connection, the standard for sound pressure has been established as a value of 0.0002 microbar by the USA Standards Institute (USASI No. A1.1-1960). The microbar is a unit of sound intensity based on a pressure of one dyne per square centimeter. Sound pressure measurements are made by specified instruments, such as the General Radio Co. type 1551-C Sound Level Meter. The loudness of an alarm device, thus, can be computed from the following general formula:

$$L_P = 20 \log \frac{P}{0.0002}$$

where L_P = sound pressure level in dB
P = sound pressure in microbars as measured for the particular alarm device
log-refers to the common system of logarithms based on the power of the number 10.

As an example, assume that the sound pressure of a certain alarm device is 2 microbars. The corresponding sound pressure level will be:

$$L_P = 20 \log \frac{P}{0.0002} = 20 \log 10,000 = 80 \text{ dB}$$

Local Alarm Power Consumption

Another important piece of information which applies to local alarms which are electrically operated is the device's electrical power consumption in watts (watts = volts x amps). Thus, if a siren draws 1 ampere at 12 volts, it consumes 12 watts of power. Another siren draws ½ ampere at 12 volts or 6 watts. Regardless of what the dB ratings might state, it is highly unlikely that the lower powered siren is louder than the higher powered one. Here again, one has to be careful in jumping to conclusions since the power drawn by electronic sirens varies with the wail rate and frequency of the siren and power figures computed by the formula may or may not be the average power. Furthermore, conversion efficiencies from electrical power consumption to acoustic power output must also be evaluated. These characteristics are determined by the siren horn structure (exponential, conical, reflex) and by the efficiency with which the siren is coupled to the air. In addition to electronic sirens there are also motor-driven mechanical sirens. These sirens are extremely loud but tend to draw very high currents, usually 10 amperes or more. Before a mechanical siren can be used in a security system, one must make certain that the power capability of the control can handle the currents required by the siren. Without making the entire subject too mysterious, it is probably true that the only meaningful way to evaluate the loudness of a local alarm is to listen to it in the environment in which it is used. If such a siren or bell is used outdoors, a suitable weatherproof enclosure must be employed and such an enclosure will also, to some extent, affect the loudness. Furthermore, the direction of the wind and air moisture, in an outdoor application, will also have an effect on the apparent loudness depending on where the listener is standing.

We have, so far, only discussed local alarm devices which are electrically powered. Since the electrical power which drives them is usually located in the control units, such a local alarm can be completely defeated by the intruder by cutting the cable which leads to the bell or siren. This condition can be prevented by encasing the local alarm cable in conduit or including batteries in the bell-siren box. These batteries would switch into the circuit automatically in the event that the wires leading to the bell or siren have been cut. Another way of overcoming this problem of local alarm defeat is to utilize a nonelectric local alarm, such as a gas driven (usually Freon) horn or siren. Such local alarm devices, in addition to being very loud, contain within the local alarm case bottled gas under high pressure. A solenoid plunger actuates the siren whenever an electrical circuit is closed or when the wires to the alarm are tampered with. In the previous section we have discussed security installations where a single local alarm in the form of a bell or siren is used. It is consequently the function of the local alarm to deter the intruder and to advise individuals in the vicinity of the premises that an intrusion is in progress. It is, of course, of equal importance to also advise remote points of the intrusion and this is accomplished through the use of regular or leased telephone lines. With regard to remote point signaling, two techniques are available: (1) the direct dialer method, and (2) the direct wire, or leased line, method.

Remote Point Signaling

Direct Dialing

In the direct dialer method, an automatic dialer (see Fig. 1-10) is connected, usually through a telephone coupler which is rented from the phone company, to an existing telephone line. Dialers usually contain a tape cartridge which is programmed to dial a series of telephone numbers in the event that the dialer is triggered. In addition to dialing the programmed numbers, the dialer also gives a prerecorded message. Many dialers are currently available having a wide spectrum of characteristics. Some use double channel tapes whereby separate calls and messages can be made in the event of fire or burglary, respectively. Furthermore, such dialers are designed to repeat their dialing sequence in the event that the number is busy when first dialed. In addition, many of these dialers have the capability of seizing telephone lines even though the line may be busy as a result of an incoming or outgoing call. Some dialers have built-in programming capability as well as offering the user the option of listening to the tape without actually dialing out. In addition, most of the dialers are designed to be used with standby power so that the dialing sequence will take place even in the event of ac power failure. In recent years, a number of units have become available employing digital dialing techniques. With this equipment, the number to be dialed is selected by the system user through a series of switches making programming extremely simple. Dialer identification is not through a voice message, but through a tone sequence which can also be preset by the user and which is then identified at the other end. Such digital dialers are physically small, contain no moving mechanical parts, and have a high degree of reliability. The only drawback that these dialers have is that there is no way for the dialer to dial out in the event that the intruder cuts the telephone lines leading from the premises. To overcome this possibility, the direct or leased wire system is used.

Direct Wire

In the direct wire technique, a private telephone line is leased from the telephone company running from the protected premises to the remote reporting point (see Fig. 1-10). This line is not a telephone line in the traditional sense, but consists of a two-wire circuit having no interface with the existing telephone equipment. At the protected premises, a special circuit called the reversing relay is employed. In this circuit, the relay is energized when the system is in its normal state. The relay is deenergized when an alarm signal is generated by the control. When the reversing relay is deenergized as a result of an alarm signal, the polarity through the direct telephone line reverses. At the remote reporting point a differential relay senses this polarity reversal

Fig. 1-10. Automatic dialer equipment for security systems.

and gives an alarm indication. Furthermore, the differential relay has three positions: positive, negative and off. The positive position (indicated by a meter or a green light) occurs when the system is normal and the line polarity has the normal configuration. In the alarm state, the polarity reverses and the differential relay now swings over to the negative position. (A red light at the remote control point gives an indication of an alarm condition.) If someone were to tamper with the system by cutting the direct telephone line, the reversing voltage through the telephone line would be lost and the relay would switch into the off position, lighting an amber trouble light at the remote control point. The amount of reversing voltage required by such a system varies anywhere from 9 to 18 volts direct current, and is dependent upon the length (resistance) of the direct telephone line.

Direct Line to Police

With regard to remote point alarm systems, whether automatic dialer or direct line, an important point with regard to programming should be made. Both the automatic dialer and the direct line can, of course, be programmed to give their alarm directly at the local police station. This technique, though used in many security installations, has received increased resistance from local police departments. The reason for this resistance is understandable. As the need and popularity of security systems increased, incoming telephone lines to police stations were literally swamped by alarm messages generated by automatic dialers. Even when special telephone numbers were assigned for automatic dialer service, system overloading could easily occur and since, regrettably, many of these automatic dialers were installed by untrained personnel, a high amount of falsing occurred. With regard to the direct line technique, a special indicator, as previously discussed, must be mounted at the police station. Here again there is a limit as to how many of these devices a local police station can physically expect to supervise with manpower at their disposal. As a result, the trend has been away, especially in the case of automatic dialers, from programming directly into the police station. As a matter of fact, many communities, by law, now forbid the programming of automatic dialers into the police station. With regard to direct lines, police stations do still accept them, but generally will restrict direct line alarm indicators to a fixed quota within their capability—assigned on a "first come first serve" basis or on a "need to have" basis (banks, jewelry stores, liquor stores, etc.). It has definitely become the trend in remote alarm reporting to program the signal, whether automatic dialing or direct line is used, into a 24-hour telephone answering service or into a central station reporting service (see Fig. 1-11). These services continuously monitor incoming alarm signals, record them, and make a logical decision as to what to do about them. This decision may include phoning a number of people, including the police, and actually sending out guards, etc. Further-

BASIC SECURITY SYSTEMS 25

Fig. 1-11. Electronic digital dialer for security systems.

more, some of the more modern central stations are equipped with computer printout systems which not only print the name and address of the premises in the alarm state, but the time of the alarm and the nature of the alarm. These central stations are equipped in such a manner that the lines coming into them can always receive alarm data even in the unlikely event that all of the alarms went off simultaneously. With regard to central stations, tremendous technological advances are currently being made in this field. Equipment is already available whereby literally thousands of subscribers are continuously scanned, via telephone lines, for violated intrusion detectors. Under such a system, the central station does not wait for the intrusion, but literally asks the subscriber if his premises have been violated. Such high speed automatic interrogation may take place as rapidly as once every few minutes.

2
Perimeter Intrusion Detectors

The most widely used method for protecting a premises against intrusion employs a simple perimeter detector. Basically, this detector is some form of electrical switch which either opens or closes as a result of the intrusion. This switch is inserted, in series or in parallel, into a wired system which controls the alarm circuitry.

Perimeter switches, in addition to being used to protect a premises against intruders, are also often employed as panic or holdup buttons in applications where the security system must be instantaneously energized in an emergency situation.

Perimeter Switches

Perimeter switches, regardless of their method of operation, have either the (A), (B) or (C) configurations which are illustrated in Fig. 2-1. In (A), the contacts are normally open, but close when energized. In (B), the contacts are normally closed, but open when energized. In (C), a contact wiper is used, which offers both opening and closing when energized. This type of switch is often referred to as a single pole, double throw (SPDT). (Refer to Fig. 1-7 for applications of these switches in alarm control systems.)

The majority of perimeter switches have contact ratings in the very low-ampere range, usually less than 50 milliamperes, although there are some exceptions. This limitation is understandable, since the currents flowing in a se-

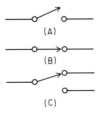

Fig. 2-1. Basic perimeter switch contact configurations. (A) One normally open contact switch. (B) One normally closed contact switch. (C) One normally open plus one normally closed contact (SPDT) switch.

curity system perimeter loop are very low. Consequently, these types of switches must never be used to directly energize devices which draw large amounts of current, such as bells and sirens. Furthermore, in evaluating these perimeter intrusion detectors, it is important to consider the possibility of contact jitter and bounce as a result of switch vibration. Such vibrations may arise due to wind gusts, machinery noise, traffic noise and other sound sources which would cause rattling in the structural member on which the switch was mounted. The contact bounce and jitter may actually open or close the contacts briefly for a period of 500 microseconds or less. If the security control circuitry does not have provisions to disregard even the briefest circuit interruption, false alarming would result due to momentary interruptions. This is especially true in control circuits using silicon-controlled rectifiers (SCR) which can respond to nano-second pulses.

A number of special installation techniques are employed in professional installations of perimeter intrusion detectors. These techniques are discussed in Chapter 9.

Magnetic Switches

The magnetic switch is the basic device used in perimeter protection circuitry. The switch consists of two units: a magnetically sensitive set of contacts and a permanent magnet. Both devices are embedded in a protective case with mounting holes. (Fig. 2-2 shows two types of magnetic switches.) In usage, one portion of the magnetic switch is attached to the fixed member, while the other is mounted to the movable member to be protected, such as a door or window. When the two parts are in close proximity (½ in. or less) the magnet energizes the contacts in the mating member and holds them open or closed as the case may be. As soon as the door is opened, the magnet is moved away from the magnetically sensitive contacts and the perimeter circuit is opened or closed, depending on which type of switch is used. A wide variety of magnetic switches are available in various sizes and capabilities. Low-priced magnetic switches tend to use weak permanent magnets which function properly only if the two pieces are in very close proximity, virtually touching each other. In the higher priced magnetic switches, powerful magnets are used in conjunction with sensitive magnetic contacts and the device will work even if the magnet and the mating are as far as 1 inch apart.

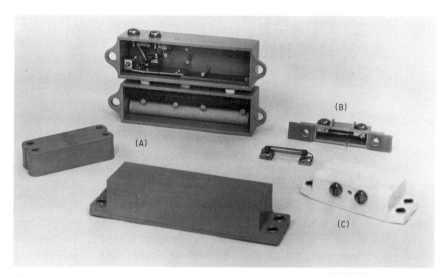

Fig. 2-2. Types of magnetic switches. (A) Spring-loaded relay-type contacts. (B) Reed-type magnetic switch (inside view). (C) Reed-type magnetic switch (assembled).

Reed Relay Contact Switches

With regard to the magnetically sensitive contacts, many switches use a reed relay contact. This type of contact consists of two overlapping, flat, ferro-magnetic reeds separated by a small air gap; one contact is fixed and the other is on a moving contact arm. The entire structure is sealed in a glass capsule which is often evacuated or filled with an inert gas to provide an oxidation-free controlled environment for the switching action. If this glass capsule with its reeds is brought into a magnetic field, the flux in the gap between the reeds will cause them to pull together. When the magnetic field is removed, the reeds will spring apart, due to spring tension within the reeds. A schematic of a typical reed switch appears in Fig. 2-3.

Reed magnetic switches sometimes have a tendency to develop a magnetic memory in the switch contacts. This condition causes the contacts to stick when the reed has been held in a magnetic field for a long period of time—a condition which is likely to occur in security applications. Furthermore, since the reeds are physically small and the contact pressure is comparatively light, reed-type magnetic switches must not be used in circuits where currents exceed a few milliamperes.

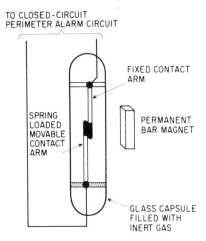

Fig. 2-3. Schematic of magnetic reed switch.

Other Magnetic Switches

Other types of magnetic switches use wiping, self-cleaning, silver contacts similar to the type used in relays. This type of magnetic switch, when used in conjunction with a good strong mating magnet, will offer very positive action, even when doors misalign and rattle momentarily as the result of wind gusts and other forms of vibration.

With regard to the electrical contact structure, most magnetic switches are of the kind where the circuit opens upon intrusion. However, a number of manufacturers produce such devices which close upon intrusion, while others have a single pole, double throw structure where either opening or closing circuits can be obtained as shown in Fig. 2-1.

As in most perimeter switch protection devices, a magnetic switch can be jumped fairly easily. In the normally closed type, all one has to do is put a jumper wire across the contacts; in the normally open variety, all that need be done is that the wires be cut. Furthermore, enterprising intruders may completely bypass magnetic switches by bringing a very powerful large magnet near the contacts from outside, thereby holding them in position while the entry is made.

Vibration Contacts

Vibration contacts consist of a fixed and a movable member with some form of contact pressure adjustment. The entire assembly is embedded in a protective plastic case. The contacts are usually of the normally closed type and the pressure adjustment is set up in such a manner that vibration caused by an intruder using a hammer, saw, crowbar or chisel will cause the circuit to open resulting in an alarm. Vibration contacts are ideally suited for mount-

PERIMETER INTRUSION DETECTORS 31

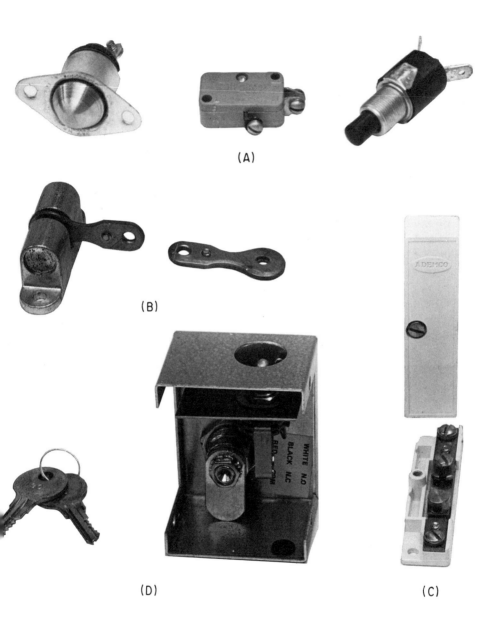

Fig. 2-4. Various perimeter transducers and panic buttons. (A) Three typical perimeter switches: (left to right) window button, microswitch, and standard pushbutton. (B) Trip wire device. (C) Vibration contact switch. (D) Panic button device with key reset.

ing on nonmoving structural members which might be subjected to vibrational action from an intruder's entry. These would include walls, ceilings, partitions and even floors. Some typical perimeter transducers are shown in Fig. 2-4.

Step Mats

A very popular perimeter detector is the step mat, which comes in a wide variety of lengths and widths. This type of detector consists of a special plastic or rubber sandwich in which contact surfaces have been embedded. As long as there is no weight on the mat, the contacts remain open. As soon as the intruder steps on the mat, the contacts close, initiating an alarm. This type of perimeter protection mat is available in extremely thin construction, down to 3/32 of an inch thick, permitting undetectable installation under carpets. Furthermore, mats are available as runners in widths as wide as 30 inches and lengths up to 25 feet. In addition, pressure-sensitive strips, pads and other configurations are also available for mounting on windowsills, underneath door jambs, on staircases, or any area where an intruder might walk. Here again, as is the case in most of the other perimeter detection devices, only a limited amount of current, less than 50 milliamperes, can be passed through such a device. Furthermore, if the mat is placed in a high density walk-in area, the contacts may wear or become excessively resistive after a year or two of usage. The detector cells in Fig. 2-5 are often used in the step mat intrusion detector shown in Fig. 1-3.

Fig. 2-5. Detector cells used as a switch with step mat.

General Purpose Contacts and Plungers

Since perimeter security detectors are basically simple switches, a wide variety of general purpose contactors exist in this generic grouping. These include: bullet switches (which may be hidden in moldings and door jambs and will electrically open or close if the window or door is opened), pushbuttons, and various forms of microswitches which will trigger if a lid, door or board is moved or opened.

Trip Wires and Traps

A number of trip wire and trap-type of perimeter protection devices are also available, and these usually consist of a spring mounted to a terminal block and a thin piece of wire with an end fitting which mates with a ball type fitting. This kind of trap is mounted across a door or entranceway at knee height or lower. When an intruder enters, he trips over the wire and pulls the wire out of the ball socket, opening the circuit and triggering the alarm. A number of variations of this basic switch are available, using not only copper wire, but metal strips, string, etc.

Foil

One of the most widely used methods of perimeter intrusion protection consists of thin metal foil which is glued, usually by means of varnish, to windows and other partitions which are likely to be broken when the intrusion is made. Obviously, this type of circuit opens upon intrusion. The foil is usually made of thin lead, having a thickness of approximately .002 inch and widths of ⅜ to 1 inch. Some foil has adhesive backing and can be fastened directly to the window or wall which is to be protected. However, the most common practice is to secure the foil to the window. This technique has the additional advantage of protecting the thin foil from scratches and disintegration.

Tilt Switches

Tilt switches are used in perimeter protection where entrances might be made by tilting an entry point, such as a transom or horizontal cellar door. Tilt switches employ a mercury filled capsule embedded in a protective plastic housing. Depending on the mounting angle, the contacts within the mercury pool will open or close as soon as the angular position of the switch is changed.

Panic Buttons

Security systems are often turned off when the protected premises are in normal usage, such as during the daytime hours in an industrial plant. However, even during the times when the security system is off, it may become necessary to instantaneously energize the alarm system in case of a holdup or some other emergency situation. In order to make this possible, a number of holdup and panic buttons are available. These devices are simple switches, the same as the perimeter type switch, but are very often hidden so that the user can activate them instantaneously in an emergency. These switches are inserted into the circuit in the same manner as perimeter switches, but may take the form of foot or knee activating devices which, in some instances, must be reset by a key.

3
Space Intrusion Detectors and Signal Processing Circuits

Space intrusion detectors, though they serve the same function as perimeter intrusion detectors—that is, the detection of an unwanted intruder—employ sophisticated electronic circuits and detect targets in three-dimensional or volumetric space. Space intrusion detectors are actually complete electronic systems in themselves and contain the detecting transducer, an amplifier, some form of signal processing, a relay driver, and a power supply or a dc regulator.

Doppler Intrusion Detectors

A number of these space intrusion detectors employ the Doppler principle which was briefly described in Chapter 1. This technique depends on the change in frequency of a signal, received by the intrusion detector, when a moving object enters the signal's path. The frequency of the received signal will be increased, for example, if the intruding object were moving towards the transmitter (the receiver and transmitter units are adjacent to each other).

This frequency change, commonly called the Doppler shift, is due to the apparent increase in the velocity of the reflected signal as received by the intrusion detector.

A simplified application of the Doppler technique is illustrated in Fig. 3-1. Transmitter "T" continuously emits ultra-high frequency (UHF) signals of frequency f-1 (assumed to be in the 900MHz range). These signals are reflected back to receiver "R" from the walls, ceiling and floor of the protected room or enclosure, as indicated by representative paths AB and CD. Now, if an intruder should enter the room, his moving body or mass would reflect signals, such as signal path EF, back to receiver "R" in less time than signal paths AB and CD. The frequency of signal path EF, thus, is increased to f-2 by the Doppler effect. This frequency shift is perceived by the space intrusion detector and the alarm is sounded. A small portion of the transmitter's output is fed directly to the receiver for comparison purposes in monitoring received signals.

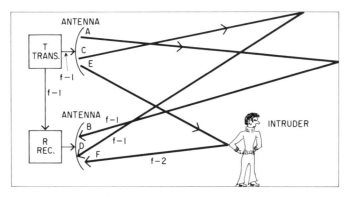

Fig. 3-1. Elements of space intrusion detecting system.

The change in frequency is determined by the velocity of the moving mass with respect to the frequency of the fixed transmitter. The amplitude of the signal is dependent upon the size and distance of the signal-reflecting target. In a typical Doppler system, energy is produced by a radio-frequency generator and this signal is beamed toward the moving object. A portion of the signal is picked off the transmitting antenna and coupled to a mixer. This signal is the Doppler reference signal and is compared with a signal returning from the moving object. A receiving antenna senses the returned signal and injects it into a mixer. The mixer combines the two signals and produces an output whose frequency is the sum and the difference between the reference signal and the signal reflected back from the target (see Fig. 3-2A). For example, at a frequency of 10GHz, each mile per hour of target velocity will produce a Doppler signal at the mixer output of 31Hz. The formula for the Doppler shift is given by:

$$f_d = \frac{2vf_r}{c}$$

where f_d = Doppler shift frequency in Hz
 v = target velocity in meters per second
 c = speed of light in meters per second
 f_r = frequency of transmitted signal.

In some Doppler devices, a separate transmitter and receiver is used while others employ a dual function oscillator. In the latter "self-excited" Doppler devices, the transmitter also serves as a receiver with the mixer directly inserted into the transmitter antenna system. More elaborate Dop-

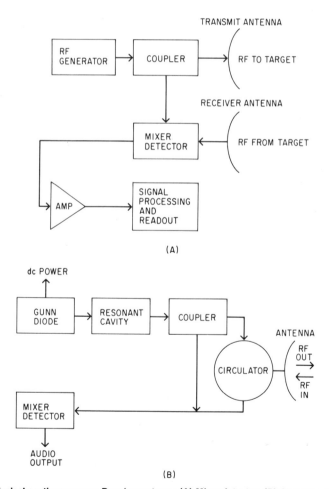

Fig. 3-2. Typical continuous wave Doppler systems: (A) Mixer detector. (B) Circulator method.

pler devices may use a circulator (see Fig. 3-2B) to isolate the transmitted signal from the received signal thereby lessening the possibility of frequency changes (pulling) which can occur from large targets near the transmitter, resulting in false alarms. A circulator consists of a magnetically biased ferrite which exhibits one-way transmission characteristics at microwave frequency and can consequently be used to separate the transmitted from the reflected signal.

In the above examples, continuous radiation of energy (continuous wave CW) is employed and such a transmission does not give range information. Sophisticated circuits are available where Doppler intrusion detectors develop range information in addition to velocity data. Such information can be extremely useful since it indicates whether an object is moving toward or away from the detector. Such data will help to differentiate between a false alarm and an intruder.

For example, a tree moving back and forth, a structural member fluctuating back and forth, a curtain, or a moving venetian blind all generate a Doppler shift. However, these objects move back and forth (toward and away) from the intrusion detector and eventually come to rest in the same spot where they started. If range data were available, one could differentiate between this type of false alarm causing-signal and an intruder who is presumed to continually walk in one direction, never returning to the same spot in a short period of time. One method which generates range data in a Doppler system utilizes a phase-comparison technique. With circuitry, the reference signal is chopped or diplexed between two closely spaced frequencies by modulating the transmitter with a square wave. A comparison of the phase difference between the received signals reflected from the target gives range data. The formula for this data is:

$$R = \frac{C}{2w1} (\emptyset 1 - \emptyset 2)$$

where R = Range
 C = speed of light
 W_1 = frequency difference of the two signals
 $(\emptyset 1 - \emptyset 2)$ = phase shift of one Doppler signal with respect to the other as long as the range is less than—

$$\frac{\pi C}{2 W_1}$$

any ambiguities in the system are resolved. If the range gains beyond the safe value, ambiguities can occur. (For example, frequency differences between the two transmitted signals of 500kHz give good range data out to 500 feet.)

With such a circuit, the switching rate must be such that the time delay of the returned signal is considerably less than the period of transmission at any one frequency. Thus, at a frequency of 10GHz, a switching rate of 100kHz is adequate for ranges of up to 500 feet. If the switching rate is too low, insufficient sampling time will be available to the system and the end result will be a loss of resolution.

The schematic diagram of a typical continuous wave 915MHz transmitter and Doppler space intrusion detector, which has a range of about a 15-foot radius, is shown in Fig. 3-3. (This device carries FCC certification as indeed must all devices of this kind operating at the frequencies of 915MHz, 2.5GHz, 10.5GHz, and 22.1GHz.)

The circuit in Fig. 3-3 consists of a modified Clapp oscillator using a PNP silicon transistor. The output is directly coupled to a one-quarter wave, ground plane antenna. The mixer detector diode is placed directly into the antenna circuit and employs a low noise Schottky (hot carrier) diode. When this transmitter is operating in free space, a steady dc voltage is developed across the diode-resistor network. If a person moves into this free space, Doppler signals will be reflected back into the antenna system and detected by the mixer. As can be seen from the formula for this type of device operating at 915MHz, the Doppler output signal for an intruder moving at a normal walking rate would be approximately 1 to 10Hz. This Doppler signal, with the dc removed by the coupling capacitor, is fed to an amplifier and signal processor which will trigger the intrusion alarm.

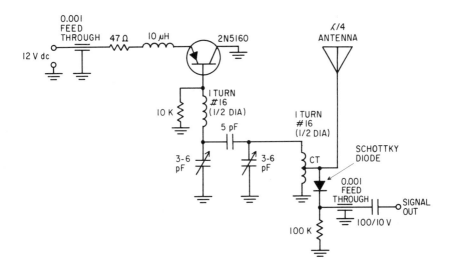

Fig. 3-3. 915MHz Doppler shift detection circuit.

Gunn Diode

The basic principles outlined above hold true for all space intrusion detectors operating at any frequency through the utilization of the Doppler principle. For example, at a frequency of 10GHz, a number of space intrusion detectors are currently available utilizing the solid state Gunn diode (see Fig. 3-4). This diode is mounted in a resonant cavity and operates at comparatively low voltages (12 volts direct current or less). Gunn diodes will emit microwave energy when the direct current is impressed across them. The diodes are highly reliable, although they are somewhat temperature sensitive and their efficiency is low, with a typical diode producing approximately 10 milliwatts of microwave energy with an input of 5 watts. See Fig. 3-5 for a noise comparison of Gunn, Klystron, and Impatt oscillators.

Fig. 3-4. Typical 10.5GHz Gunn diode assemblies (pencil included for size comparison): (A) Microwave Associates unit. (B) Hewlett-Packard unit. (C) General Electric device. (D) Amperex Gunn diode connected to horn antenna.

The Gunn diode oscillator, just like discrete transistor oscillators, may use a circulator to separate the transmitted signal from the reflected signal. Self-excited Gunn diode oscillators can also be used with the Doppler signal taken off directly from the diode itself. In these cases, the Gunn diode acts not only as an oscillator, but also as a mixer. A schematic of a Gunn oscillator, amplifier, and power supply is shown in Fig. 3-6. Fig. 3-7 illustrates the physical size of the power supply and amplifier for the Gunn circuit.

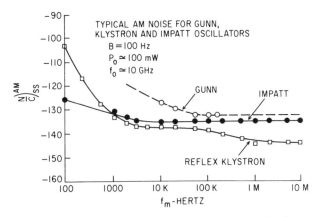

Fig. 3-5. Noise comparisons for Gunn, Klystron, and Impatt Oscillators.

Other Solid State Diodes

In addition to the Gunn diode, a number of other solid state signal sources are currently available, each operating in the 10GHz region. Below 10GHz, that is, at 2.5GHz and at 915MHz, discrete oscillators are usually used. One of these devices is the bulk-effect diode, utilizing the transferred electron effect for direct energy coverage from a dc power source to microwave energy. Another solid state device which is currently being used for the generation of microwave energy is the Impatt diode, which depends on a reverse-biased pn junction in order to generate microwave frequencies. The efficiences of the Impatt diode are as low as that of the Gunn diode, but higher powers at lower noise figures can be generated. For example, Impatt diodes are available which can generate up to one watt of power at 10GHz. However, the Impatt diode, unlike the Gunn diode, requires a comparatively high voltage, between 70 and 100 volts, and these voltages are not usually available in security systems unless a dc to dc converter is used, adding expense to the system. As a result, Impatt diodes have enjoyed little popularity in security systems. (Prior to the advent of solid state systems, Klystron or tube oscillators were usually used for the generation of high frequencies.)

FCC Regulations Regarding Space Intrusion Detectors

This material is taken from FCC Docket 13863 of August 18, 1971, pages 119 and 120.

§ 15.303 *Restriction on operation*

No field disturbance sensor may be operated unless it has been certificated and labeled as complying with the requirements of this part.

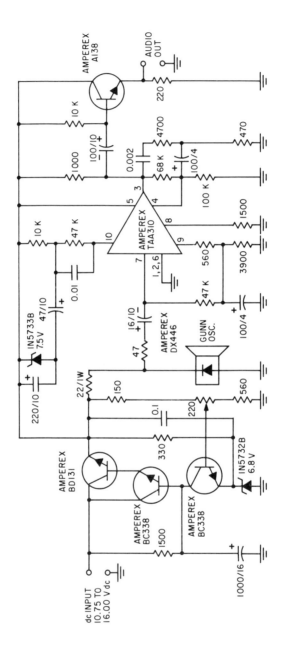

Fig. 3-6. Schematic of Gunn power supply, amplifier and oscillator circuit using Amperex Gunn diode.

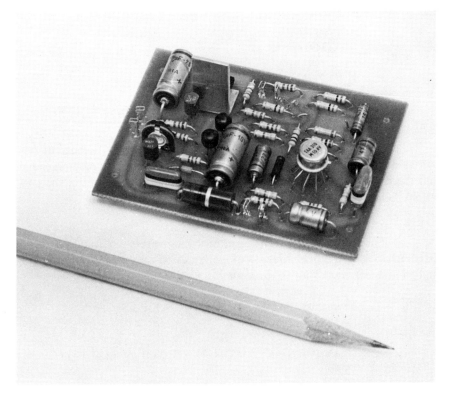

Fig. 3-7. Power supply and amplifier for Gunn diode circuit (pencil shows size comparison).

§ 15.305 *General technical specification*

(a) A field disturbance sensor may be operated on any frequency (including frequencies above 900MHz) subject to the requirement that the field strength of emissions on the fundamental or on a harmonic or on other spurious frequencies shall not exceed 15 μV/m at a distance of $\lambda/2\pi$ from the sensor. (The distance $\lambda/2\pi$ is equivalent in feet to 157 divided by the frequency in MHz.)

(b) Alternative to the above, a field disturbance sensor may be operated on any frequency listed as follows (subject to the technical requirements set out in §§ 15.307 and 15.309): 915MHz, 2,450MHz, 5,800MHz, 10,525MHz, and 22,125MHz.

§ 15.307 *Permitted bands of operation*

The carrier frequency of a field disturbance sensor operating on one of the frequencies listed in § 15.305 and any modulation components thereof shall be kept within the following band limits:

Nominal Operating Frequency (MHz)	Band Limits (MHz)
915	± 13
2,450	± 15
5,800	± 15
10,525	± 25
22,125	± 50

To minimize the possibility of out-of-band operation because of frequency drift due to aging of components or other causes, it is recommended that the carrier frequency be kept within the central 80 percent of the permitted band.

§ 15.309 *Emission limitations*

(a) For a field disturbance sensor operating within any frequency band listed in § 15.307, the field strength of emissions on the fundamental shall be limited in accordance with the following:

Frequency (MHz)	Field Strength
915	
2,450	50,000 μV/m at 100 ft
5,800	
10,525	
22,125	250,000 μV/m at 100 ft

(b) Spurious emissions (including emissions on a harmonic of any frequency listed above, shall be suppressed at least 50 dB below the level of the fundamental; however, suppression below 15 μV/m at 100 ft is not required.

For pulsed operation, measured field strength shall be determined from the averaged absolute voltage during a 0.1 second interval when field strength is at its maximum value. Below 1,000MHz, the measurement bandwidth shall comply with the requirements set out in the American National Standards Institute Specifications C63.2−1963 and C63.3−1964. Above 1,000MHz the measurement bandwidth shall be 5MHz.

§ 15.311 *Interference from a field disturbance sensor*

(a) Operation of a field disturbance sensor is subject to the general conditions of operation set out in § 15.3.

(b) The operator of a field disturbance sensor who is advised that his sensor is causing interference to an authorized radio service shall promptly stop operating the sensor, and operation shall not be resumed until the condition causing the harmful interference has been eliminated.

§ 15.313 *Certification of a field disturbance sensor*

The procedure for certification of a field disturbance sensor is identical to the procedure for a radio control for a door opener as set out in §§ 15.260−15.264 inclusive, and §§ 15.268−15.272 inclusive, except that § 15.264(b) shall not apply.

§ 15.315 *Description of measurement procedure*

The report of measurements shall describe in detail the measurement procedure that was used. If a published standard was used, reference to the standard is sufficient, provided any departure from the standard is described in detail.

§15.317 *Frequency range over which measurements are required*

(a) For a field disturbance sensor operating below 100MHz, the spectrum shall be scanned from the lowest frequency generated in the device up to 1,000MHz. Field strength for all significant emissions shall be measured and reported.

(b) For a field disturbance sensor operating above 100MHz the spectrum shall be scanned from the lowest frequency generated in the device up to 10GHz, provided that for sensors operating on frequencies above 5GHz, the spectrum shall be scanned to the highest frequency feasible, above 10GHz. Field strengths of all significant emissions shall be measured and reported.

Antenna Radiation Patterns and Characteristics

With regard to antennas, Table 3-1 shows some of the radiation patterns and propagation characteristics which can be expected from various space intrusion detectors operating in different modes. Thus, in accordance with electromagnetic theory, as the frequency goes higher, the radiations begin to take on more and more the characteristics of light. Consequently, with devices operating in the 10.5GHz region, extremely sharp beams can be generated through the utilization of microwave horn antennas. Furthermore, as a general rule, antenna gain will vary with antenna beam width. For example, more forward range can be expected from a horn antenna with a beam width of 30° than could be expected with a beam width of 60°. Horn antennas with various beam widths are available which offer anywhere from as little as 8 dB to as much as 32 dB of gain. The net result, bearing in mind the power limitations set by the FCC, is that a high gain, narrow beam width antenna may give a forward coverage, at 10.5GHz, of as much as 300 feet. At 2.5GHz, the patterns become less sharp, while they are virtually omnidirectional at 915MHz where highly directional antennas are not practical due to size.

Table 3–1. Antenna Radiation Patterns and Propagation Characteristics

Operational Mode	Antenna Pattern	Energy Passes Through:	Energy Reflected By:
915MHz	Omni-directional	Brick, glass, concrete, wood plaster.	Any metal, foil insulation, wire lath, screens.
2.5GHz	Beam-shaped	With some loss through brick, glass, concrete, wood, plaster.	Any metal, foil insulation, wire lath, screens, as well as some optically opaque substances.
10.5 and 22.1GHz	Beam-shaped	Only very thin sections, with losses, of optically clear material like glass.	Any metal, foil insulation, wire lath, screens, as well as virtually all optically opaque substances.
Sound	Omni-directional	Nothing (only cloth, etc.)	Any solid substance
Ultrasonic	Omni-directional	Nothing (only cloth, etc.)	Any solid substance
Infrared	Omni-directional or straight line	Nothing (only cloth, etc.)	Any solid substance
Visible light	Omni-directional or straight line	Optically clear materials (glass, etc.)	Any optically opaque substance

With regard to losses, 10GHz signals are blocked, for the most part, by all structural members, including brick, cinder block, wood panels and, of course, all metal including foil insulation and wire lath. This characteristic becomes less pronounced at 2.5GHz, and at 915GHz the energy easily passes through virtually all materials except metal.

With regard to space intrusion detectors which operate in other modes—sonic, ultrasonic, visible and infrared light—similar propagation rules apply. Thus, optical devices, whether invisible or infrared, can be focused in extremely sharp beams, while ultrasonic energy is highly diffused. It should be pointed out that these latter radiation modes are not subject to FCC control. The distances which can be covered are normally dependent upon the transmitted power, transmission losses in the air or other medium, and the receiver's sensitivity.

Space intrusion detectors operating in the Doppler mode at frequencies 915MHz, 2.5Hz, 10.5GHz and 22.1GHz require extremely stable power supplies since any fluctuation in the power might be interpreted as an intrusion signal by the amplifying circuit. The requirement of a very stiff power supply cannot be over-emphasized, especially for Doppler devices operating in the 915MHz region where the detection of a walking intruder delivers very low frequency Doppler signals on the order of 1 to 10 hertz. Obviously, any momentary transient or change in power supply regulation could be interpreted as an alarm signal by the subsequent amplification system. The noise generated by the oscillators themselves, as Table 3-1 shows, must be carefully evaluated to ascertain whether the noise spectra in the Doppler frequency of interest is higher in amplitude than the signal which will be reflected back. In addition, very careful power supply design must be employed since regulation circuits, especially Zener diodes, often produce troublesome noise spectra.

Furthermore, devices operating at 915MHz might be affected and falsely trigger by strong, harmonically related, spurious radiations such as might be generated by radio transmitters operating in the 450MHz region. Such signals might cause false alarms unless sophisticated signal processing circuitry is employed. These problems are less pronounced in the 2.5GHz, 10GHz, and 22.1GHz bands where spurious signals are less likely to occur. In addition, 915MHz energy will propagate through nonmetallic walls and structural members and might falsely trigger on an object moving outside of the protected premises. This property can sometimes be used to great advantage if the 915MHz space intrusion detector is required to give security protection for two floors. As previously indicated, the higher radio frequencies usually do not go through partitions and walls.

Sonic and Ultrasonic Space Intrusion Detectors

The Doppler space intrusion detection principle will also operate with ultrasonic and sonic signals. (With this type of energy, FCC certification is not required and any amount of transmitted power can be employed.) The transmitters in ultrasonic devices generally employ crystals made of barium titanate and operate at frequencies from 15 to 40kHz. In sonic Doppler space intrusion detectors, modified loudspeakers are used to emit the transmitted audible frequency which is usually between 4,000 to 8,000 hertz. The mechanism which generates the Doppler signal with ultrasonic and sonic intrusion detectors is the same as in the case for electro-magnetic (radio-frequency) devices. However, in these intrusion detectors, a separate transmitter and receiver is invariably used with the "mixing" action taking place in the receiver. Fig. 3-8 shows a sonic Doppler intrusion alarm which operates at 3,000Hz and contains an internal siren. The Doppler signals which are produced, both in terms of frequency and amplitude, are essentially of the same order of magnitude as is produced by RF devices, and a considerable amount of amplification and signal processing is consequently required.

Most ultrasonic intrusion detectors employ the continuous wave Doppler technique. Here again, as is the case with radio-frequency devices, some imaginative methods have been employed to increase reliability and lessen the susceptibility to false alarms. In some of these ultrasonic transducers, the emitted signal in the ultrasonic range is actually frequency modulated, and the phase differences between the emitted and received signal are detected. Furthermore, since a multiplicity of these devices may often be used to cover large areas, great care is taken to insure the frequency stability of each device.

For example, let us take the case where two ultrasonic detectors overlap in their range pattern. The first of these operates at 35,100Hz, while the second operates at 35,200Hz. If the frequency of either device drifts through

Fig. 3-8. Sonoguard portable intrusion alarm unit.

the other, a beat signal in the Doppler range will be generated. This beat signal will, of course, appear as a false intrusion signal. The likelihood of such a frequency drift in moderately-priced ultrasonic devices is quite high. In order to guard against this, manufacturers of these devices often cite the operational frequency of their ultrasonic unit and suggest to the user that the widest guard band speed be employed in the event that a number of devices are used together. Thus, the likelihood of a 40kHz ultrasonic space intrusion detector drifting through a 30kHz ultrasonic space intrusion detector is quite low, and a considerable margin of safety is inherent in such a system when two are used together with overlapping patterns. To achieve an even higher degree of stability, some detectors employ crystal controlled, temperature compensated oscillators.

A typical master control unit used with the Aerospace Research Ultrasonic space intrusion detector is shown in Fig. 3-9. This is a sophisticated device which employs elaborate signal processing circuits. It can differentiate between objects which move only in one direction (and exhibit a continual change in range) and objects which move back and forth that show no change in range.

Amplifiers

The amplifiers which are used to process Doppler signals must be extremely stable and have very good signal-to-noise ratios, since they must

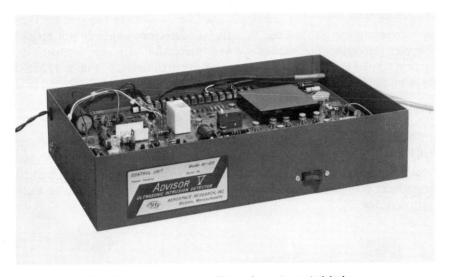

Fig. 3-9. Aerospace Research Ultrasonic master control device.

handle signals in the microvolt region. Furthermore, in order to achieve stable, clean amplification, the bandpass characteristics for such amplifiers are limited to the frequencies which are of interest. Thus, an amplifier used in a 10GHz space intrusion detector would be designed to amplify microvolt signals in the 10 to 100 hertz region—a typical Doppler signal for an intruder normally walking in a 10GHz field. It should be remembered that amplifiers, even with narrow bandpass characteristics designed to operate in the microvolt region and having large amounts of gain—up to 90 dB are not easily designed and built. This is especially true for amplifiers which operate at the very low frequencies such as 1 to 20 hertz, where $\frac{1}{f}$ noise in discrete transistors, IC's and operational amplifiers increases in a nonlinear fashion. A number of IC chips are currently available which offer, on a single chip, up to 120 dB of gain, but if the data is carefully interpreted one can see that these gain figures are usually applicable only at RF frequencies and not at the audio Doppler frequencies which are commonly encountered in intrusion detector circuits. Operation amplifiers, some of which have excellent noise figures, have the problem of requiring a balanced power supply which is usually not available in the typical security system. Furthermore, even with current state of the art devices, neither IC's or operational amplifiers can compare—with respect to $\frac{1}{f}$ noise—to the best discrete transistors designed for small signals, low noise, and high beta. Consequently, Doppler signal amplifiers usually contain discrete stages of preamplification followed by bandpass amplifiers which often employ IC's or operational amplifiers, since the signal levels are much higher.

Low level, high gain, bandpass amplifiers which are required for Doppler detectors are also needed in intrusion detectors which do not employ Doppler principles. For example, a very successful method of intrusion detection employs a silicon chip measuring approximately 1/8 in. x 1/2 in. x 1/16 in. thick which has a nominal resistance of 1,000 ohms.

Stress-Sensitive Transducers

Figure 3-10 depicts the details of a typical stress-sensitive transducer. The silicon chip is shown cemented to a bar in A. In B, this unit has been molded into a plastic container with the wires attached to it. The completed transducer is shown attached to a wooden structure in C for use as an intrusion detector.

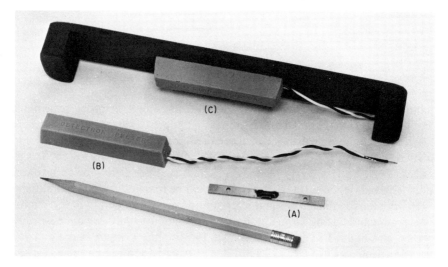

Fig. 3-10. Stress-sensitive transducer for intrusion detector (pencil shows size comparison).

When such a chip is deformed by slight flexing, its internal resistance will change, even if the deformation is as little as .001 inch. If this chip is epoxied to a beam or other structural member which will deform if an intruder walks upon it, a definite detectable resistive change will take place. A number of stress detectors can be connected in a bridge driven by a very stable dc source. Several of these transducers may be connected in parallel in a half-bridge circuit as shown in Fig. 3-11. A single balance potentiometer is used to balance the entire network and a very stable 10 volt supply is used to drive the bridge. The 10k resistors act to compensate the circuit for temperature fluctuations which would cause changes in the stress transducer resistance. A small resistive change which is the result of an intruder walking upon the stress transducer will cause a fluctuating signal across E in Fig. 3-11. If a

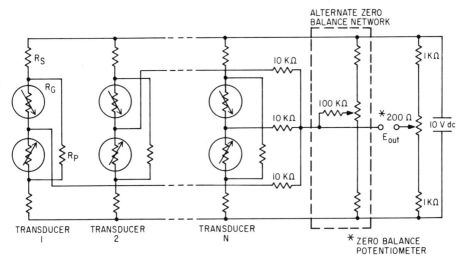

Fig. 3-11. Resistor bridge connection for stress-sensitive transducers.

coupling capacitor is inserted into the output, the dc will be blocked, and a signal will result whose frequency is dependent upon the intruder's walk rate, normally 1 to 10Hz. The amplitude of such a signal is dependent upon the amount of flexure which the chip sees as a result of the intruder's weight, and the rigidity of the structural member involved. A typical single transducer, depending upon the structural quality of the beam to which it is epoxied, can cover as much as 50 to 100 square feet, and obviously a number of transducers are capable of giving volumetric intrusion protection to very large areas. Furthermore, since the frequency of the phenomena is roughly equal to the signals produced by a Doppler device operating at 915MHz, a similar kind of amplifier can be used.

Seismic Transducers

Another space intrusion detector which operates in a similar mode as the stress detector is the seismic transducer, which contains a piezoelectric crystal encapsulated in a protective sheath as illustrated in Fig. 3-12. These seismic detectors, sometimes called geophones, are usually buried in a shallow depth of 4 to 8 inches. Soil conditions must be such that good coupling is attained since seismic waves have difficulty in passing through frozen or extremely hard ground. With respect to a human intruder, typical coverage for such a detector is a circular pattern having a 25-foot to 200-foot radius (an area of 1,965 to 125,660 square feet). If protection against vehicles is desired, circular areas having a radius of 1,500 feet (7,000,000 square feet) can be attained.

52 ELECTRONIC SECURITY SYSTEMS

Fig. 3-12. Teledyne Geotech seismic intrusion detector.

Audio Detection Intrusion-Detectors

Another popular type of space intrusion detector operates on the audio-detection principle and does not use the Doppler mode. These detectors employ a series of special microphones which are strategically located throughout the premises to be protected.

Existing public address systems with the speaker turned into a microphone while the premises are under protection are often used for intrusion de-

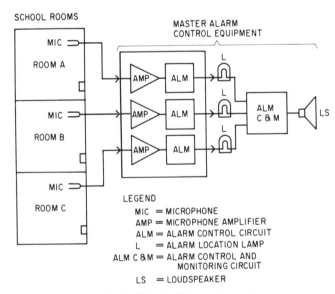

Fig. 3-13. Audio-detection intruder alarm system.

tection. This is one of the reasons why audio-detection devices are very popular in schools where existing public address speakers terminate in every room.

The audio-detection intrusion detector listens to the sounds in the protected area, and triggers an alarm relay if an intrusion takes place.

The block diagram of a simple type of an audio-detection intrusion alarm system is depicted in Fig. 3-13. The audio-detection intrusion alarm system is applicable, for example, in schools for detecting intruders after school hours. A microphone is concealed in the classrooms and connected to an amplifier in the master alarm control equipment. The outputs of the individual audio amplifiers are adjusted to prevent the activation of the alarm on the normal or ambient room noises when the rooms are unoccupied. The sound caused by an intruder would be amplified sufficiently to trigger the alarm circuit. A lamp could be utilized to indicate the particular room or location of the microphone that was activated. The security guards also could monitor the alarm system to check for possible false alarms.

Differential Technique

A number of unusual signal processing circuits must be employed in these devices for the purpose of eliminating the possibility of false alarming. One processing technique employs a two-input differential amplifier. The first input, input A (see Fig. 3-14), accepts the signal which the audio detection device receives from the microphones within the protected area. The second signal, input B, is taken from a microphone which is placed outside the protected premises.

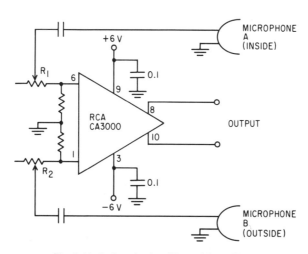

Fig. 3-14. Audio-rejection differential amplifier.

Let us assume that such an installation is in a building located in a high traffic noise area. The prime difference between the signal at input A and the signal at input B is the amplitude of the noise signal. Obviously, the amplitude at B would be larger than at A, due to the fact that the B microphone was outside, near the traffic noise, while the A microphones were inside, where traffic noises were attenuated as a result of walls, etc. Other than a difference in amplitude, both signals at A and B are about the same, with regard to their acoustic spectra. (There may, of course, be some minor nonlinear differences, due to the fact that certain frequencies are attenuated more than others by structural members such as walls, but these differences are negligible and can, for all intents and purposes, be disregarded.)

Consequently, if in such a circuit, potentiometer R_2 in the B leg is adjusted to attenuate the B signal, so that both signals A and B are at the same level at the input of the differential amplifier, the differential amplifier will then cancel signals which are the same level and will amplify only the difference between any two signals. This cancelling capability is usually called the amplifier's "common-mode rejection" capability, and is extremely high with modern solid-state devices. In the example cited above, once signals A and B are equal, the differential amplifier will not have any output, regardless of the amplitude or spectral content of the noise.

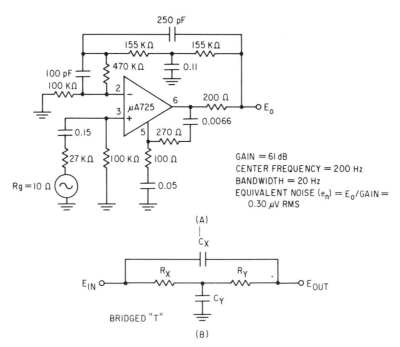

Fig. 3-15. Schematic of active bandpass filter for intrusion alarm systems.

It becomes apparent from such a circuit that if an intruder enters one of the protected rooms and commences to make noises or talk with his accomplice, this signal would only be heard by the A microphone and not by the B microphone, which is placed outside. The differential amplifier would immediately amplify the signal resulting in an alarm condition.

Audio-Frequency Filters

In addition to the differential technique, audio-detection devices have further sophisticated processing in the form of complex electronic audio filters such as are shown in Fig. 3-15. The human voice, indeed virtually any recognizable sound, has a distinct acoustic spectrum. For example, the male voice generates a unique frequency spectrum, with pronounced peaks in the 500 to 200 hertz region. On the other hand, a dog barking, a radiator hissing, a bell ringing, would generate signals with a different acoustic spectrum. The audio signal developed by the microphones placed within the protected premises can be processed through a series of audio-frequency filters, which are tuned to let signals of certain frequencies go through, while rejecting signals which are not in the area of interest. Thus, through the careful design and selection of such bandpass filters, differences can be recognized between human intrusion signals and noises which are of no interest with regard to security purposes.

Photoelectric Intrusion Detectors

In addition to space detectors which depend upon sonic, ultrasonic, or microwave radiations, a number of very effective devices exist which operate on photoelectric principles. The simplest of these is a visible light system

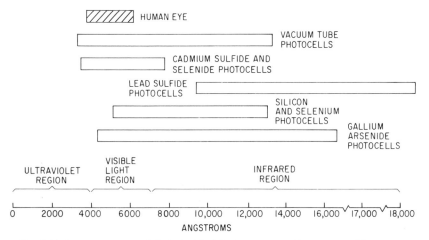

Fig. 3-16. Response curves of photoelectric detectors compared to response of human eye.

whereby an incandescent light source, through optics, aims a beam at a photoelectric detector such as a silicon or a cadmium sulfide cell. (Silicon will generate a voltage when light falls upon it, while cadmium sulfide changes in resistance when light falls upon it.) These photoelectric cells are terminated in an amplifier which triggers the alarm whenever the light beam is interrupted. Visible light photoelectric intrusion detectors have a number of disadvantages, in that the beam can be seen by the intruder and the device essentially works over straight line paths, even though complex patterns can be formed through the utilization of mirrors. With regard to the visibility of the beam, photoelectric devices are available which employ an infrared beam that cannot be seen. The spectral response of the human eye is compared to the response of typical detectors in Fig. 3-16.

Invisible infrared beams also can be generated through the utilization of standard incandescent light bulbs, interposing optical filters which remove the visible components while letting the infrared energy go through. In recent years, another excellent source of infrared energy has been widely used in these intrusion detectors. This source is a semiconducting diode made from gallium arsenide, which is a strong infrared transmitter in the 9,000 angstrom region. In addition to emitting infrared energy, gallium arsenide diodes can actually be pulsed in the laser mode. A number of outdoor laser infrared intrusion systems are available and these devices produce an extremely narrow invisible beam which is effective up to 5,000 feet or more.

In infrared systems, as in visible light beam systems, careful beam alignment of the transmitter and the receiver must be achieved. Furthermore, in infrared systems, it is often the practice to modulate, or chop, the emitted signal, so that it becomes difficult to bypass the system by holding an infrared source, such as a burning cigarette, in front of the receiver. An even more sophisticated infrared system, which does not require a separate infrared transmitter and receiver, employs a diffraction grating which is placed in front of a gallium arsenide diode infrared detector. The diffraction grating actually consists of a piece of infrared transparent material, into which a series of lines are etched. When this detector looks into a three-dimensional environment, it will no doubt see a number of false alarm infrared sources, such as pipes, radiators, light bulbs, etc. However, all of the infrared energy which falls on the diode through the diffraction grating remains at the same amplitude, since the sources are not moving. An intruder now enters this three-dimensional space, and he becomes a new source of infrared energy. The wave length of human infrared radiation is approximately 9½ microns. (This energy corresponds to approximately 90° F.) A considerable amount of human infrared radiating signal is retained, even as the body heat signal passes through clothing and even if the intruder enters a cool environment, which would tend to lower his body temperature. As the infrared emitting intruder moves through the protected area, his infrared emissions actually sweep across the diffraction grating and are chopped by the lines in the diffraction grating, resulting in a group of pulsating "interferometer" waves. These waves appear as signal pulses at the diode with a frequency depending upon the speed at which the intruder moves and the number of lines in the diffraction grating. These pulses are taken from the diode and are amplified in subsquent circuitry.

Still another passive photoelectric device employs a number of photoelectric cells in a balanced bridge circuit with each cell viewing a portion of the three-dimensional environment. When an intruder enters, the energy (whether infrared or visible light) falling on one of the cells will be different than on the other three, causing a momentary bridge imbalance. The energy difference at any one of the cells is due to the fact that the intruder casts a shadow on one of the cells, or interrupts the light energy path to the cell. The resulting imbalance in the bridge circuit is detected as a momentary pulse across the bridge and is amplified for purposes of alarm triggering.

False Alarms

As we have previously indicated, one of the major problems in security systems is the possibility of false alarming. In the very simple perimeter transducer systems (magnetic switches, foil, etc.) the likelihood of a false

Table 3−2. Effect of False Alarms on Different Systems

False Alarm Causing Phenomena	Will tend to falsely trigger:				
	RF	Ultrasonic	Sonic	Infrared	Stress
Lightning	YES	NO	NO	NO	NO
Power Line failure transients	YES	NO	NO	NO	NO
Power line switching transients	YES	NO	NO	NO	NO
AC sparks from switches and contacts	YES	NO	NO	NO	NO
RF transmitters (cabs, police cars, citizen's band, etc.)	YES	YES	NO	NO	NO
Hot and cold air currents	NO	YES	NO	NO	NO
Randomly rotating machinery (fans, etc.)	YES	YES	YES	NO	NO
Randomly moving objects (chandeliers, venetian blinds, curtains, trees)	YES	YES	YES	NO	NO
Small animals (birds, dogs, cats, rodents)	YES	YES	YES	YES	NO
Noises (telephone bells, radiator hiss, etc.)	NO	YES	YES	NO	NO
Structural member flexure due to heat, cold, wind	NO	NO	NO	NO	YES
Wavering in large metal surfaces (walls, roofs, air and heating ducts)	YES	YES	YES	NO	NO
Random heat sources (incandescent bulbs, radiators, sunlight)	NO	NO	NO	YES	NO

alarm is somewhat less than in systems utilizing sophisticated space detectors. On the other hand, perimeter detectors are easily bypassed and tend to offer little protection. Volumetric space intrusion detectors are very difficult to bypass and employ sophisticated circuitry and techniques. As shown, the initial signals which these space intrusion detectors develop (whether they be Doppler, ultrasonic, optical, stress, etc.) are quite small—in the microvolt region—requiring considerable amplification. It becomes obvious that the sensitive amplifiers used in such equipment could possibly be triggered by conditions other than an intruder entering the protected premises, thereby causing a false alarm. The phenomena which can cause a false alarm are numerous, and reference to Table 3-2 gives some indication of what can be expected. It is possible, through conservative design and the utilization of only the best components, to bring the sensitive amplifiers used in space intrusion detectors to a point where they will be as stable as current state-of-the-art will permit. However, in order to drastically reduce and even eliminate the possibility of false alarming, additional signal processing circuits are often employed.

Signal Processing Circuits

One of the most commonly used signal processing circuits is the Schmitt trigger. The circuit shown in Fig. 3-17 constitutes not only a basic Schmitt trigger, but also functions as an events counter and an integrator. The circuitry, including transistors Q_1 and Q_2, is such that any given input signal exceeding a certain amplitude will generate a square wave pulse of fixed amplitude. Thus the output of a Schmitt trigger is always a square wave pulse of fixed amplitude and pulse width, regardless of the wave shape of the input signal. Consequently, the input signal to a Schmitt trigger can be virtually any shape, including transient pulses, sine waves, triangular waves, complex waves, etc. The only requirement to trigger the Schmitt circuit is that the input amplitude reach a certain value before the Schmitt fires. In the circuit shown, R1 is a potentiometer which precedes the input to the Schmitt and selects the amount of input signal which can be fed to the Schmitt. In this circuit, the Schmitt will trigger with a 200-millivolt signal. Thus, R1 acts as a basic sensitivity control which can be set to fire the Schmitt circuit on various levels of input signal. Once the Schmitt circuit has fired, it produces a square wave pulse which is rectified by diodes D_1 and D_2, so that all pulses become positive. These positive pulses are then fed into an events counting-integration network consisting of R_2, R_3, and C_1. As one can see from the circuit, capacitor C_1 is charged up at a rate depending upon the number and amplitude of the trigger pulses. The amplitude of these pulses to C_1 is set by R_2. Once the charge on C_1 reaches a certain level, transistor Q_3 fires and this transistor in turn fires transistor Q_4 in whose collector circuit the alarm relay

SPACE INTRUSION DETECTORS AND SIGNAL PROCESSING CIRCUITS 59

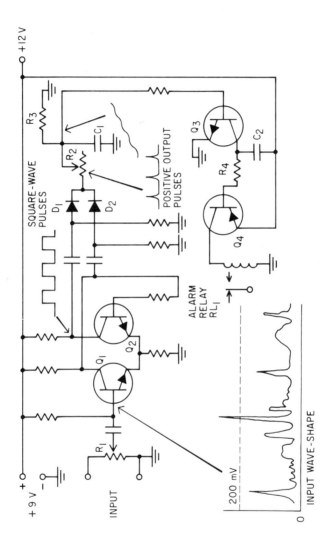

Fig. 3-17. Schematic of Schmitt trigger circuit.

RL1 is connected. (The purpose of the network consisting of R4 and C2 is only to hold the alarm relay and its associated driver transistor in for a short period of time, thereby eliminating transients.) Getting back to the event counting-integration circuit, once capacitor C_1 has fired transistor Q3, current flows and C_1 is discharged. (Any small charge remaining on C_1 is bled off by resistor R_3.) After this has happened, the counting-integration process can start all over. As one can see, the potentiometer R_2 actually acts as an events counter. If R_2 is at minimum resistance, every pulse coming from the Schmitt trigger, but also functions as an events counter and an integrator. other value, only a portion of the pulse coming from the Schmitt circuit will be applied to capacitor C_1. Thus, the number of pulses required to trigger the alarm relay can be set by varying the control R_2.

This circuit has some extremely useful signal processing characteristics. In the first place, a large signal at the input of the Schmitt circuit counts for no more than a small signal. As long as the Schmitt triggering voltage is reached, in this case 200 millivolts, the circuit will fire once. Consequently, very large false alarm signals such as lightning or other transients, trigger the Schmitt circuit only once, just as would an intruder. However, the event-counting potentiometer R_2, further on down in the circuit, would disregard the transient if it were set up so that approximately three pulses from the Schmitt circuit were required to bring C_1 up to charge, firing the next transistor Q_3.

Still another excellent method of signal processing employs digital methods. As previously seen, in the Schmitt technique, events are actually counted by charging up a capacitor. Regrettably, every capacitor has leakage and as soon as an event pulse is introduced it begins to leak off. Leakage in capacitors is comparatively high, and if events occur spaced a few seconds apart, the capacitor in the Schmitt circuit cannot store indefinitely and will consequently never trigger the alarm, due to the fact that the capacitor leakage drains the event faster than the Schmitt circuit can deliver. This problem is overcome in a circuit which stores every event as long as desired and will trigger an alarm only if a preset number of events in a preset time period occurs. This method employs digital processing techniques of the binary system such as used in computers, data processing equipments, and electronic telephone switching systems. The schematic diagram in Fig. 3-18 illustrates a simple pulse-counting shift register stage, such as used in digital processing. Explanations of the related binary system operations are outside the scope of this book.

The input to the digital processor can come from a Schmitt trigger. However, instead of charging a capacitor, the signal is fed to the shift register (also called a flip-flop circuit) which counts and stores each event. In the circuit shown in Fig. 3-18, every time an event is entered, the next flip-flop transistor triggers which fires an associated buffer amplifier such as Q3. The buf-

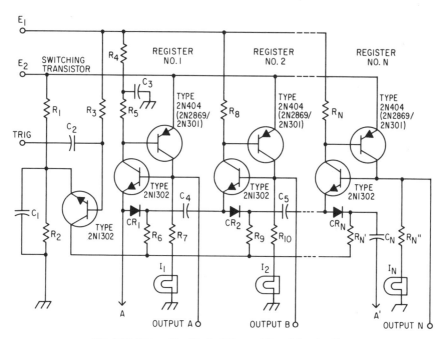

Fig. 3-18. Schematic of typical binary shift register circuit.

fer amplifier lights a lamp indicating how many events have taken place. A great number of these counters can be staggered after each other. A selector switch is introduced into the circuit and is set to a given event. (The selector switch has the alarm relay in its wiper circuit, while the contacts of the selector switch are simply wired across the various event lights.) If the selector switch is set to event 8, as soon as the eighth event light is energized an alarm will be initiated. In addition to having infinite event counting capability, such a circuit can also store events for very long periods. Furthermore, events can be entered serially (one after the other in time) or even simultaneously. A digital circuit which has the capability of accepting events simultaneously contains an electronic delay circuit at its input stage. Thus, if two events arrive simultaneously, the second event is delayed a few microseconds at the input with the end result that the two simultaneous events actually give a two-event count.

In addition to selecting the number of events in a digital processor, it is also necessary to have an adjustable time sample base. Fig. 3-19 shows a digital time sampling processor. This time sampling processor, shown in the diagram in Fig. 3-20, is adjusted to reset the digital processor from a second to a minute after the first event has taken place. For example, in typical usage,

62   ELECTRONIC SECURITY SYSTEMS

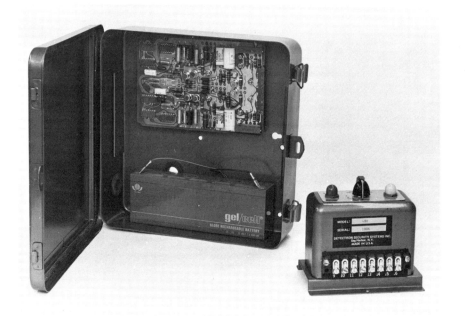

Fig. 3-19. Detectron #485 digital time sampling signal processor.

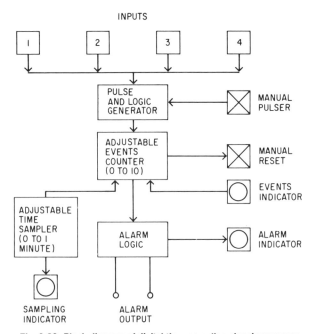

Fig. 3-20. Block diagram of digital time-sampling signal processor.

such a digital processor may be set up to trigger the alarm circuit if four events take place in a 10-second period. In the case where only three events take place, the digital processor would automatically reset itself and start counting once again.

4
Security Controls

In the previous chapters, we have seen that perimeter and space intrusion detectors result in a contact closure or opening when violated. It is the function of the security control to take this closure or opening and process it in such a way that an alarm condition will result. Since many individual intrusion detectors are usually employed throughout the premises, wireless (radio) techniques can be used to transmit the signal back to the control. A number of wireless systems are available and these will be discussed in a separate chapter. However, the most widely used technique employs hard wiring from the intrusion detector back to the control. This method, though more expensive to install, offers the highest degree of reliability and the lowest false alarm rate.

Hard Wiring Security Circuits

The simplest hard wired security circuit consists of a battery and an alarm indicating device such as a bell with the intrusion detector closure in the alarm state. As illustrated in Fig. 4-1A, such a circuit will cause the bell to ring whenever the circuit is closed. This basic circuit has a number of drawbacks which make it, in the practical sense, virtually useless for security purposes. In the first place, the bell would only ring as long as the intrusion detector was violated. Thus, someone entering very rapidly through a door protected by a magnetic switch would cause the bell to ring for a second or two at best. Furthermore, as can be seen from the circuit, all that need be done to defeat the system would be to cut the wire, thereby permanently holding the

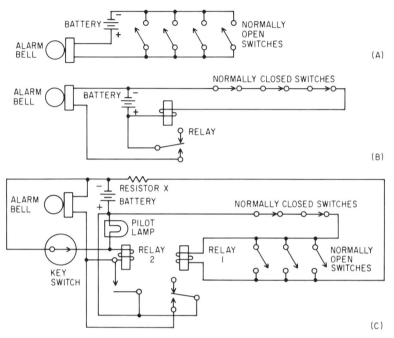

Fig. 4-1. Basic security circuit.

circuit open. In addition, this circuit is limited to intrusion detectors in which the circuit closes in the alarm state. A slightly more complex circuit (Fig 4-1B) employs a relay which is connected in series with intrusion detectors which open when violated. Here the alarm bell would sound not only when the detector was violated, but also whenever the wiring was cut, thereby offering a certain amount of tamper protection. Once again, this circuit is not very practical since the bell will ring only as long as the detector remains open. Furthermore, any detector can be bypassed merely by putting a jumper wire across it. Although versions of both of these circuits are sometimes used in low priced security controls, the majority of better controls employ a circuit which overcomes the many disadvantages of the two simpler techniques (Fig 4-1C). This circuit is referred to as the two-wire, supervised, perimeter loop circuit and exhibits the following characteristics:

1. Cutting through any one of the circuit wires will energize the alarm.
2. Shorting one wire to the other will energize the alarm.
3. Intrusion detectors which close a circuit as well as intrusion detectors which open the circuit can be used and intermixed.
4. The alarm circuit, once energized, remains in the alarm condition even though the circuit is restored to normalcy. The only way that the alarm circuit can be reset is at a key operated reset switch.

Let us examine this circuit in greater detail. Relay 1 is the alarm detection relay. Note that this relay energizes when the perimeter loop is in its normal condition. If any normally open intrusion detector closes, it places a short across the two perimeter wires which in turn short circuits the relay coil. Consequently, the relay opens and an alarm condition results. Resistor X in this circuit controls the additional power which must be absorbed due to the fact that relay 1 is now shorted. If this resistor was not in a circuit, a dead short would result across the battery, and it would drain itself very rapidly. If any normally closed intrusion detector opens, it interrupts the current to the relay coil and the relay once again opens, sounding an alarm. Once relay 1 has opened—even momentarily—it energizes relay 2 which is normally in the deenergized position. The operated contact on relay 2 is used to power the alarm bell, and to latch or hold this relay operated even though relay 1 releases because the perimeter loop has been restored to normal. Thus, the alarm continues to ring, once energized, and can only be reset by releasing relay 2 which is accomplished by momentarily opening its latching or holding circuit through the key switch. This circuit, and variations thereof, is the circuit which is most widely used in security systems. Generally speaking, alarm relay 1 is a sensitive relay which requires a very little power (a few milliwatts) for its operation. The reason for this is obvious. It would be poor engineering to send large currents through the perimeter loop. (Such large currents would only be required if alarm relay 1 were a heavy-duty type.) In order to carry large voltages and currents through the perimeter loop, heavy wire would be required and the intrusion detector switches would also have to handle heavy currents, making them more expensive. Furthermore, as the circuit indicates, the sensitive relay is normally energized. Thus, it draws power and should do so with the least amount in order to give the standby batteries maximum life in the event of power failure. A wide variety of these sensitive relays exist and they are designed to operate at various voltages and currents. A typical perimeter loop system may employ a sensitive relay operating at 6 or 12 volts dc operating on currents from as little as 3 milliamperes to 10 milliamperes. Even with the use of sensitive alarm relays operating at very low power, there is usually a limit to the total amount of perimeter loop wire which can be used. Manufacturers of security systems state this limit in maximum permissible resistance of the perimeter loop expressed in ohms. For example, a typical security control might indicate that the maximum perimeter loop resistance must not exceed 500 ohms. By looking at a table of typical wire resistances per 1,000 feet, one can properly compute the kind and length of wire which should be used:

Wire Gauge (Solid Copper)	Resistance in ohms per 1,000 ft
24	26
22	16
20	10
18	6

For example, No. 24 solid copper wire has a resistance of 26 ohms per 1,000 feet. If it is used in the two wire loop system, the total footage has to be doubled for the loop run since two wires are used. Thus, a two wire perimeter loop which goes through a plant and uses up 2,000 feet of two-conductor No. 24 cable would have a total resistance of 4 x 26 or 104 ohms. This is within the limit of a security system where the maximum permissible loop resistance is 500 ohms. The value of 500 ohms is only representative. Some controls can take a total loop resistance of 1,000 ohms while others may not operate properly if the loop resistance is above 100 ohms. The upper limit of the loop resistance is, of course, dependent upon the sensitivity of the relay and its operating voltage. As the circuit shows, the loop voltage is in series with such a relay coil and the loop resistance will result in a voltage drop. If the loop resistance becomes too high, the drop will be too great to energize the relay. Furthermore, if the relay draws excessive currents the standby battery life will be diminished appreciably. Such a condition is highly undesirable, especially with security controls which do not have an ac power supply and so rely solely on batteries for operation. In this type of control, batteries will have to be replaced periodically since the circuit does draw a finite amount of current. However, good circuit design dictates that the batteries associated with the system should be replaced about every 6 months.

Silicon Controlled Rectifiers (SCR)

The circuits which we have discussed so far utilize relays to accomplish the various alarm functions. These functions can also be accomplished through the use of silicon controlled rectifiers (SCR's). These solid state devices can switch large loads with very small signals and furthermore will latch up once they have been put into the alarm state. A number of SCR circuits are used in security controls using intrusion detectors which open or close when violated. Although SCR type security controls have become increasingly more popular in recent years, they are somewhat prone to false triggering since any transient induced in the wiring can cause the SCR to fire. Such transients can be inductively picked up from power lines or radiating machinery. Heavy filtering and bypassing must consequently be employed in an SCR control to eliminate the possibility of false triggering.

In the circuit shown in Fig. 4-2, a mixture of normally open and normally closed intrusion detectors may be used. The capacitor C 1 serves to filter any transients which might occur in the circuit. The SCR is triggered by the transistor driver and directly actuates the alarm bell. The amount of current drawn by the alarm bell is limited only by the current handling capabilities of the SCR which in this case is 2 amperes. SCRs handling much larger currents, up to 10 amperes and more, are currently available. Once the SCR has triggered, the alarm will stay latched until the reset key switch is momen-

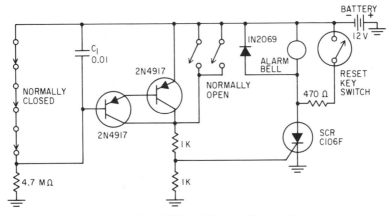

Fig. 4-2. Schematic of SCR circuit for security control systems.

tarily actuated, thereby unlatching the SCR and rearming the control. This particular circuit does not contain a system status indicator which could, of course, easily be added.

Security Controls

Locations

It is generally good practice in security installations to locate the control in some out of the way place away from general view. The reasons for such placement are obvious in that it discourages tampering by unauthorized personnel. Typical placements include closets, basements, and locked cabinets. In placing the security control, it must be remembered that the user will have to get at it for operational purposes. Furthermore, with controls which have ac power supplies, 120 volts normally would be fed to the control. It has become practice, in the case of ac powered controls, for the manufacturer to supply an outboard transformer which plugs directly into an ac outlet. The low voltage output of this transformer is then wired back to the control eliminating the problem of high voltage ac wiring.

Operation

Virtually all controls contain on their front panel a key operated malfunction switch which controls and tests for the following functions (Fig. 4-3):
1. System On (Armed)—System Off
2. Bell or Siren Test
3. Standby Power Test

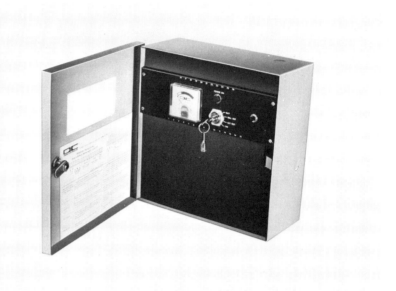

Fig. 4-3. Security control box equipped with circuit status meter and test switch.

With regard to the system "on/off" position, this circuit function is often divided into two conditions called "day" and "night." When the switch is in the "day" position the normal alarm portion of the circuitry is bypassed so that the associated bells and sirens do not sound while protected areas are entered during the normal course of business. However, the supervised aspect of the loop circuit remains active so that if anyone cuts the wiring during the day—in preparation of an unauthorized entry at night—a separate alarm indicator (buzzer or light) will be immediately energized. Furthermore, many controls contain provisions for a holdup-panic circuit. Such a circuit is terminated at strategic positions throughout the premises utilizing some form of pushbutton. When an emergency occurs, for example, a holdup, such a button is depressed immediately energizing the alarm system. Obviously such a circuit must remain operative even while the control is in the "day" position. When this switch is in the "night" position, the system becomes fully armed and operates as a complete security system.

The "bell" or "siren" test position is used in making certain that the alarm bell or siren remains functional. These alarm indicating devices are often mounted outdoors and contain mechanical parts which wear. Thus, they are subject to malfunctioning and freezeup, a condition which might never be discovered unless they are tested each time the system is ready to be armed.

The third position of the switch tests the condition of the standby battery or the condition of the batteries in those controls which are entirely battery operated. These batteries may be of the ordinary dry cell variety or, in ac powered controls, may consist of rechargeable batteries continuously charged by the power supply. In the event of power failures these batteries are automatically switched into the circuitry and consequently the condition of the standby power must also be periodically tested to be sure it is at full charge so that the system will function in the event of power failure. When the control switch is placed into this test position the standby power is often switched directly to the alarm device, such as the bell or siren, thereby dynamically testing the standby power supply. In some controls, the best test and standby power test position are combined. If in this position the bell does not sound, the fault may be with the standby power or the bell and the system has to be checked at these two critical points. The status of the standby power is also often monitored with a meter which is placed across the standby power circuit when the switch is in this test position. As a matter of fact, meters are often employed not only to indicate the condition of the standby power but also to continuously show the integrity of the wiring and associated intrusion detectors. For example, take the case of a security installation where a protected window has been inadvertently left open. The system which was in the "day" position is now placed in the "night" (armed) position, and will immediately go into the alarm state since the open window circuit appears as an intrusion. In order to overcome this problem, security controls usually have some indicator which shows the status and integrity of the circuit at all times. For example, an indicator light may be employed which remains energized as long as all intrusion detectors are in their normal state. This indicator light will go out if any transducer is violated, regardless of the setting of the key switch. In such a circuit, the user would not go from the "day" (unarmed) to the "night" (armed) position if he saw that the status light was extinguished, warning him that a detector is in a violated state either because of an opening which has not been closed or because someone was still left in the building. The indicator light technique has the drawback that the bulb in such a system can burn out. Consequently, a number of security controls employ a meter which shows the integrity of the protective circuitry.

Key Switches

Up to this point we have limited our discussion to the functions of the control switch and indicators used directly on the control itself. In a number of applications, additional control flexibility is, of course, required. Let us take the example where a control is located on the inside of a protected premises and is armed with a key switch in the front panel. Once it is armed, the system becomes active and the user must leave through an exit which is not part of the protected circuitry. Obviously, if he leaves through a protected exit, he will immediately energize the alarm. This situation can be overcome through the use of a second special switch which is called a shunt lock. This type of switch is mounted in an exit door and is wired across the intrusion detector which is used at that exit point, for example, a magnetic switch. As long as this shunt lock is energized, the door can be opened and closed without the system being violated. As soon as the user leaves through the shunt lock wired exit door, he opens the shunt lock with his key and the exit now becomes armed. A number of variations of this shunt lock concept are available and these include a special trip mechanism which permits the user to open and close a protected door once without energizing the security system. When the door is opened a second time—presumably by an unauthorized person—the security system will trigger.

Remote Control Stations

Another method to overcome the exit problem is through the use of remote control stations. Such remote control stations serve, in a limited way, the same function as the main key switch station. Remote control stations which are either key or pushbutton operated (Fig. 4-4) are located at various

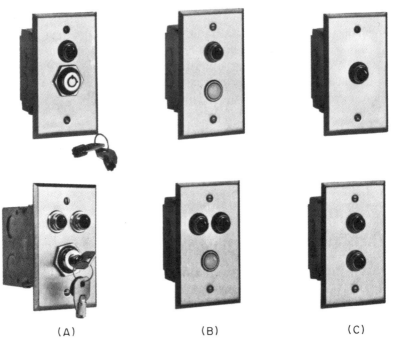

Fig. 4-4. Typical remote key switch and alarm lamp stations. (A) Round key types. (B) Pushbutton types. (C) Indicator stations (show status of alarm system).

strategic points, including exits and entrances, throughout the protected area. These stations can be used to energize and de-energize the system and give a constant indication of the status of the system. Thus, a remote control station could be located outside the protected area and can, consequently, be used to energize the entire system without triggering the system upon departure. Furthermore, remote stations become very useful when a system is to be energized internally. For example, in the case of a home, the owner may wish to energize the security system upon retiring. Consequently, a remote station is placed near his bed and may be used to protect the premises while he is asleep. This type of remote station usually contains pushbuttons rather than key switches since it is impractical to carry a key around for this type of application. It is important to point out that security controls are usually designed to handle a virtually infinite number of remote control stations. Any individual station can be used to energize or deenergize the system and, furthermore, every station shows whether the system is ready to be armed (in other words, all intrusion detectors are GO) or is in effect armed. In order to accomplish this condition some form of memory is required back at the control and this is usually accomplished by means of a mechanically latching stepping relay.

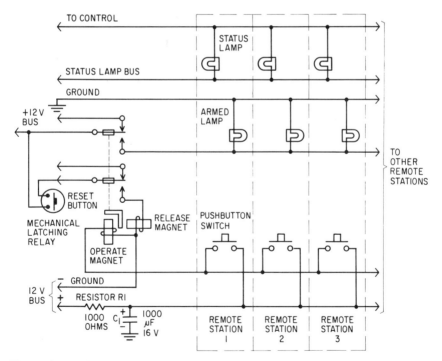

Fig. 4-5. Schematic of remote key switch stations and mechanical latching relay for control.

A number of circuits are employed and the one shown in Fig. 4-5 is representative. In this circuit condenser C-1 is continually charged through resistor R-1 which is connected to the control's 12 volt bus. Any number of remote control stations can be inserted into this circuit with each station containing two indicator lights and a momentary pushbutton or key switch. If any of the pushbuttons or key switches are depressed, the condenser C-1 discharges through the operate magnet of the energized relay. This type of circuit has the advantage of drawing very little current even though as much as 800 mA is required to energize the relay. This is due to the fact that the energy stored in the condenser, which is charged in a fraction of a second, is discharged through the relay winding. Consequently, the current in this circuit remains minimal even if one of the key switches is held depressed continuously. In such a case, the resistor R-1 limits the total current flowing in the circuit to under 12 mA. The impulse latching relay itself is of a double pole, double throw (DPDT) type.

One section of the relay energizes and de-energizes the security control. Other contacts of the relay serve to light the armed light indicators on the remote control stations. The status light circuitry is brought out directly from the control and at all times indicates the integrity of the security system thereby preventing arming of the system in the event of a violation.

The mechanical latching relay releases when the reset button is pressed. This operation sends a current pulse from the control circuit to energize the release magnet of the latching relay. The latching relay, is thereby, returned to its released position.

Electronic Time Delay Mechanism

Although the shunt lock technique or the externally mounted remote control station system overcomes the problem of arming the system and then leaving through a protected exit, they have one drawback. The sophisticated intruder can actually see the shunt lock or the remote control station and, if he is at all skillful, can pick the lock and disarm the system. The remote key switch stations are usually protected with a tamper mechanism so that the system will trigger in the event someone attempts to remove the switch cover plate. However, this does not prevent the lock from being picked, thereby bypassing the entire system.

A technique which overcomes this problem has become increasingly popular in the more professional controls and this technique uses an electronic time delay (see Fig. 4-6), which can be set by the user, giving him a fixed amount of time to leave the premises via a protected area without triggering the system. Once this time delay has elapsed (the user has now left) the system arms itself automatically. With controls having this capability, it now

SECURITY CONTROLS 75

Fig. 4-6. Detectron #LR-10 Electronic Latching Relay Control Unit.

becomes possible to locate any remote key switch station within the protected premises. The exit delay problem also comes into play when the user re-enters the premises. In systems using shunt locks or externally located remote key switch stations, he must remember to de-energize the system before entering. If he forgets to do so, the security system will immediately trigger. If trip type shunt locks are used, the false triggering of the system is inevitable since this type of shunt lock only permits a single violation without tripping the system. This violation occurred upon departure and the security system will obviously trigger upon re-entry.

Here again, the re-entry problem has been overcome in professional security controls by means of an electronic delay. A typical electronic time delay circuit is shown by the schematic diagrams in Fig. 4-7A and B. The time delay may be adjusted from a fraction of a second to 10 to 20 seconds by the installation of specified capacitors and resistors in accordance with the chart in Fig. 4-7C. The capacitor and resistor arrangement controls the time for firing the silicon-controlled rectifier (SCR) in the delay circuit which is designated Q13. Such controls, in addition to having an adjustable electronic exit delay also have an adjustable re-entry delay. In such a circuit, the system becomes armed after the preset exit delay has expired. If an intrusion detector in such a system is violated, a new circuit comes into play which delays the tripping of the alarm circuitry by an adjustable amount of time. Thus, when

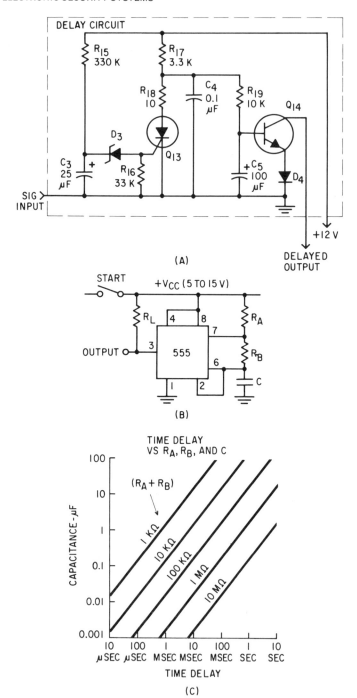

Fig. 4-7. Typical time-delay electronic circuits.

the user re-enters the premises, he has a preset amount of time to get to the control station to disarm the system. If he fails to disarm the system within the preset period, the alarm will ring. This technique has the advantage that it will virtually prevent lock picking by the intruder. Such an intruder upon entering would start the countdown cycle. Even if the intruder knew or saw the remote control station he would only have a short time period to pick the lock before the system fired. Exit and re-entry delays in controls containing this feature are usually adjustable from a few seconds to approximately 1 minute. Even if they are set to the maximum delay period it is highly unlikely that the intruder is sufficiently skilled to enter the premises, find the control station, and pick it in the 1 minute which is allowed to him.

Regardless of what technique is used with respect to the exit and entry problem, if the user completely forgets the fact that his premises are protected the alarm will, of course, sound instantaneously or after the delay periods have expired. To overcome this unlikely, though possible situation, a number of controls employ a double alarm circuit. The circuit is designed in such a way that an initial alarm will ring instantaneously whenever an intrusion detector is violated, regardless of the setting of the delay circuitry. This alarm, often referred to as the first alarm, is usually in the form of a discreet buzzer which probably would not be heard by the intruder, but would be heard by the user reminding him that the inevitable cycle has commenced. The user immediately remembers that he must get to a control station and turn off the system before the alloted delay runs out. Once the delay runs out, a second alarm closes and this alarm energizes the bell, siren, and other intrusion indicating devices.

Automatic Recycling Controls

Another feature which is very useful is the control's automatic recycling capability. The circuits of controls having this feature are designed in such a way that the entire system will automatically rearm itself, after having been violated, providing that the violated intrusion detector returns to its normal condition. For example, let us take the case of a security system whose owner cannot be reached. The premises protected by the system have been violated by an intruder who enters through a door which is protected by a special mat. The alarm sounds and the intruder is frightened off even before the police arrive. If the system did not have an automatic recycling feature, the alarm would continue to ring until the owner disarmed the equipment with a key switch. This would regrettably be the case even though the intruder is now gone and the police have arrived. If, for example, the police are unable to reach the owner and not having a key switch, they might possibly damage the system in an attempt to silence it. In controls containing automatic recycling, the system would remain in the alarm state for 5 or 10 minutes—sufficient time to frighten the intruder and summon help—and then automatically reset itself as long as all the intrusion detectors are normal. Of course, in some applications, the recycling feature is deemed not to be desirable and controls having this capability usually offer a choice between permanent alarm latch up and recycling.

Zone Selection Circuitry

It also is often necessary for security applications to have the capability of selecting and controlling various security zones. Thus, a typical building may have four floors, each one of which has been wired for security with the circuits being brought back to the control individually. It may become necessary to arm the first three floors while the fourth floor remains unarmed due

Fig. 4-8. Design Control #12-066 security control station.

to the fact that a night shift is active on that floor. Consequently, the security control which is used in such an installation must offer zone selection circuitry in the form of pushbuttons (see Fig. 4-8) or selector switches. A number of controls are currently available which have this feature.

Control Terminal Strips

The wide variety of capabilities and features which are currently available in security controls must be carefully analyzed in order to select the best control for each application. In summary, a typical control terminal strip is shown in Fig. 4-9 which represents the circuitry which might be used in a

SECURITY CONTROLS 79

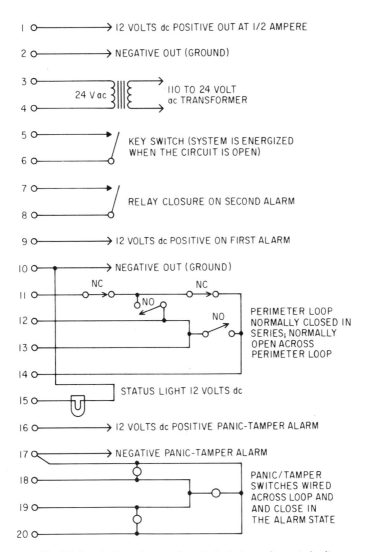

Fig. 4-9. Terminal board connections for typical security control unit.

80 ELECTRONIC SECURITY SYSTEMS

typical security installation using a single key switch station. Note that this particular control offers 12 volts dc at ½ ampere for the purpose of powering auxiliary devices such as bells, sirens, indicator lights, etc. (Chapter 5 discusses the matter of power supply capabilities in greater detail.) The input of the control consists of 24 volts ac which is furnished by means of an outboard transformer. A single pole, single throw key switch is used to energize this control and the circuit is arranged in such a manner that the control is energized when the key switch is electrically opened. This circuit feature has the interesting advantage that the control is automatically energized if someone were to cut the wires going to the key switch. This particular control uses a double alarm configuration, with the first alarm being used as a local warning in addition to serving as a dynamic test of the control. The second alarm, which gives a relay closure, may then be used to operate the bell or the siren. The control uses the standard four-wire supervised loop wiring (terminals 11 through 14) in which normally open and normally closed intrusion detectors may be intermixed (see Fig. 4-10). The system status light, which always

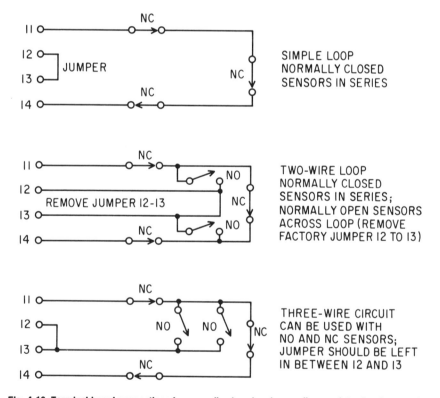

Fig. 4-10. Terminal board connections for normally closed and normally open intrusion detector devices.

shows the integrity of the wiring, connects from ground to terminal 15. Furthermore, the control does contain a panic circuit which is always armed regardless of the key switch position. Once the panic button is depressed, a panic signal of 12 volts dc appears across terminals 16 and 17. Note that in this particular control the panic circuit can also be used as a daytime tamper protection circuit since the circuit will fire either when the alarm is shorted or cut. In this control the maximum resistance recommended for the perimeter loop is 300 ohms and a number of techniques for wiring intrusion detectors into the perimeter loop are also shown.

5
Power Supplies and Standby Power

Most modern security systems are low voltage devices which can be used with virtually any type of cable. Operating voltages for these security systems vary from 1½ volts to 24 volts depending upon the voltage requirements of the accessories which are used with the system. As previous chapters have indicated, some security controls rely solely on batteries for their operating power, while other systems employ ac power from the 120-volt line which is then converted into dc power for control operations. Furthermore, these ac powered controls often have standby battery systems consisting of ordinary dry cells or rechargeable batteries which are continuously kept charged by the power supply. Such batteries are automatically switched into the system in the event of ac power failure. The importance of the security system's power supply cannot be overemphasized and its capabilities must be understood and carefully studied before an installation is made. The following specifications should be analyzed with respect to ac operated controls:

1. What is the maximum continuous and intermittent power output of the supply?
2. What protection does the power supply contain with respect to line and other transients?
3. What is the ripple content of the power supply under full load?
4. What is the regulation of the output voltage as the load changes from none to full; what is the regulation of the output voltage as the input line voltage varies?

The following specifications should be understood with respect to standby batteries and battery power supplies:
1. What is the shelf life of the battery?
2. What are the volt-ampere and discharge characteristics of the battery?
3. In the case of rechargeable batteries, what type of recharging circuitry is used?

With respect to power output of any supply, whether ac operated or battery, an initial system wattage consumption calculation must be made based on the usual formula, watts = volts × amperes. This calculation is even applicable to the low voltage ac transformer which is usually supplied with a security control system. This type of transformer plugs directly into the ac outlet and lowers the 120 volts to the required ac input voltage. Consequently, small size wiring can be used without the need of conforming to the electrical codes. Transformers of this type have wattage or volt-ampere ratings and these must be known to make certain that the security system's power requirements on an intermittent and a continuous basis do not exceed the capabilities of the transformer. If the capabilities of the transformer are exceeded, there is a good chance that it will burn out or that large voltage drops will develop which will make the entire security system unstable.

Continuous Power Consumption

The term "continuous power consumption" means the amount of total power, alternating or direct current, that the security system requires on a continuous basis while in the normal operating condition. This is actually the power consumed by the control circuitry itself, indicator lights, and the power required by various intrusion detectors. Intermittent power requirements are based on the system's power consumption in the alarm state, i.e., with all bells, sirens, lights, etc. going. The actual power calculations (W = VA) can be made at dc levels, i.e., at the battery, or after the power supply, or at ac levels right at the transformer input.

Tables 5-1 and 5-2 give some typical power consumption requirements which can be expected from various 12-volt devices used in a security system. The tables included show the kind of power requirements which may be encountered in a typical 12-volt dc security system. This system consumes 11.4 watts in the continuous mode and 39.92 watts in the alarm mode. Choosing the proper power supply for this system and allowing for a margin of safety, it becomes apparent that approximately 1 ampere is required in the alarm mode, all at 12 volts dc. These figures can be extrapolated back to the ac transformer.

TABLE 5-1.
TYPICAL POWER REQUIREMENTS FOR 12 VOLT DC DEVICES

Device	Approximate DC (Amperes)	DC Power (Watts)
Incandescent indicator light	0.080	0.96
Indicator panel meter	0.001	0.012
Security control circuitry	0.050	0.600
Ultrasonic intrusion detector	0.030	0.360
Infrared intrusion detector	0.150	1.80
Microwave intrusion detector	0.350	4.20
Perimeter intrusion detector	none	none
Local alarm (Mallory Sonalert)	0.010	0.120
6" bell	0.250	3.00
10" bell	0.400	4.800
Low power electronic siren	0.350	4.20
High power electronic siren	1.50	18.0
Mechanical siren	10.0	120.0
Standard telephone dialer with tape running	0.750	9.0
Digital telephone dialer running	0.100	1.20

TABLE 5-2.
TYPICAL CURRENT DRAWN BY 12 VOLT DC DEVICES

Item	Approximate Continuous Current (Amperes)	Approximate Alarm Current (Amperes)
Control circuitry	0.050	0.050
Three remote stations with dual lights	0.600	0.300*
Ten perimeter detectors	none	none
Four ultrasonic detectors	0.300	0.300
Sonalert local alarm (first alarm)	none	0.010
High power electronic siren (second alarm)	none	1.5
Telephone dialer	none	0.750
Total direct current (amperes)	0.950	2.910
Power consumed (watts)	11.4	39.92

*In the alarm state the system status light goes out.

In security controls using electronically regulated supplies, a voltage drop can be expected through the regulation circuitry. This drop is usually in the neighborhood of 4 volts. Consequently, to generate 12 volts dc, the electronically regulated supply has to be fed with a minimum of 16 volts ac from the transformer. If the ac voltage drops below 16 volts ac the regulator will still absorb 4 volts and, consequently, will be unable to deliver a 12-volt output. If the voltage rises above 16 volts ac the regulator will, of course, absorb the excess because the voltage regulator is within the ± 4 volt regulation range.

So, good design would call for an 18 to 20 volt transformer to take care of possible line voltage variations. The current consumed by the electronic circuitry has already been taken into consideration and, consequently, in the alarm state the transformer involved would have to deliver (ideally speaking) 18 volts at 4 amperes or 72 watts. If the security system employs devices which operate on voltages other than 12 volts dc, different power calculations will, of course, have to be made. It should be remembered that, for a given output, a device operating at 6 volts dc will draw more current than a device operating at 12 volts dc. The wattage consumed by each device will be the same. Consequently, if a power supply has a rating of 12 volts at 5 amperes, its output power is 60 watts. However, such a power supply cannot be used with a device which draws 6 volts at 10 amperes, which is also 60 watts. One might initially assume that such a device would work with a 50 percent loss in intensity, in the case of a light bulb for example. In point of fact, what would happen is that such a load might operate at 50 percent intensity for a few seconds and then would cause the power supply to burn out since 10 amperes were flowing through a circuit rated at 5 amperes, even though the total wattage is within the prescribed limit.

Transients and Noise

Transients and noise are also generated by the energizing or de-energizing of inductive devices, such as motors, transformers, machinery, etc. (see Fig. 5-1). Such inductive devices may be located considerable distances from the security control and yet the transients which they develop will travel down the power line and appear right at the security control input. In addition to transients of this type, power lines are usually noisy. The origin of this noise is sometimes difficult to establish and may be due to distant lightning, changes in line loading, or switch gear operation at the power station. Noise can also be caused by devices which have interrupting contacts, such as buzzers and bells. Consequently, troublesome noise may be caused within the security system itself when, for example, a bell is operating.

The suppression of transients and noise is of the utmost importance in a security power supply, especially if this power supply is used to power, further on down in the circuit, sensitive sophisticated space intrusion detectors, such as ultrasonic, microwave, or infrared devices. These devices usually contain sensitive solid-state amplifiers which, as a previous chapter has shown, have

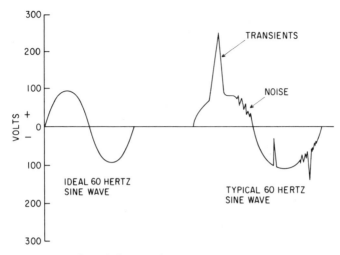

Fig. 5-1. Graph of sine waves and transients.

bandpass characteristics usually in the 1 to 100 Hz region. Transients and noise in the power line would be immediately detected by these amplifiers and no matter how sophisticated the signal processing might be, it is quite likely that a false alarm will result. The suppression of such transients consequently becomes a matter of prime importance when sophisticated intrusion detectors are used. In the event only perimeter intrusion detectors are employed, the transient problem is less severe but cannot be entirely ignored since the control circuitry itself very often contains transistors which are sensitive to such signals. This is especially true for transistorized circuits such as SCR (silicon controlled rectifier) circuits and transistorized time delay and logic circuits.

Regrettably, line transients and noise cannot be entirely eliminated, especially if they have large amplitudes and very narrow pulse widths. However, modern security power supplies contain a number of innovations which tend to limit these unwanted phenomena. Of prime importance is sound internal circuit layout and good grounding techniques. Designers of power supplies take great care to keep transient carrying circuits away from sensitive devices. Furthermore, all ground points should be terminated individually to a common ground thereby avoiding troublesome ground loops. In instances where transient problems exist, grounding the entire system to a good external ground, such as a water pipe, may prove helpful although in some instances can generate more problems than it eliminates. With respect to grounding, the entire technique is at best empirical so that a trial and error approach usually is employed. Other forms of transient and noise elimination include the use of special diodes which are placed across the power supply input at the earliest possible point, often in the ac transformer primary (Fig.

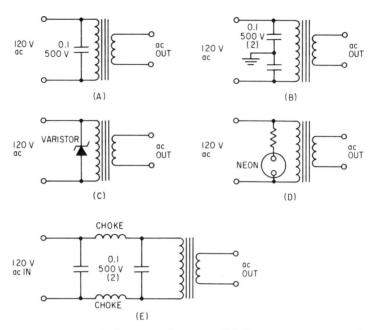

Fig. 5-2. Typical transient and noise suppression circuits. (A) Simple capacitor bypass. (B) Dual capacitor bypass to ground. (C) Diode suppression. (D) Neon tube suppression. (E) Low current filter network.

5-2). These diodes have a high resistance at the normal circuit voltage. They virtually become a dead short at transient voltages which are higher than normal, thereby momentarily short-circuiting the transformer. Such diodes are extremely fast, can catch very rapid transients, and are able to absorb considerable amounts of power for short periods of time. These diodes are commercially available under a variety of trade names, including Varistors, Stabistors, etc. Still another method of effectively reducing transients and noise is through the use of special filter networks consisting of capacitors in combination with inductors (chokes) which are inserted in critical lines. Regrettably, this technique tends to be too sluggish to detect and eliminate rapid transients. Finally, the modern integrated voltage regulators which are often used in power supplies contain special sensing circuitry with Zener diodes which are designed to clamp down and suppress even the most rapid of transients.

Ripple, Line Regulation, and Load Regulation

In an analysis of power supplies, it is also important to know the device's specifications with regard to ripple, line regulation, and load regulation. Ripple is the amount of 60 cycle alternating current which is superimposed or the direct current delivered by the supply. This property is expressed as a percentage of the dc voltage and invariably increases as the power supply load increases. It is also sometimes expressed in terms of millivolts ripple on the dc voltage. The very best electronic supplies may have ripple less than 5 millivolts under full load. With respect to regulation, the power supply specifications should include output voltage variations from no-load to full-load and output voltage variations as the input line voltage varies. Here again, parameters are expressed in percentages or voltages. Thus, the best electronically regulated power supplies may have load regulations where the output voltage does not vary more than 50 millivolts from no-load to full-load at any given voltage. Furthermore, they may have as little as 25 millivolts change in the output voltage as the ac input voltage varies from 100 to 125 volts ac.

Of course, the problems of ripple and regulation are minimal in a security system employing only batteries as a power source. A battery obviously has no ripple and its regulation, although there are some fluctuations with load, is excellent when fully charged. Of course, a majority of modern security controls employ ac power supplies exclusively or in conjunction with battery standby. The simplest ac power supply is the half-wave type which con-

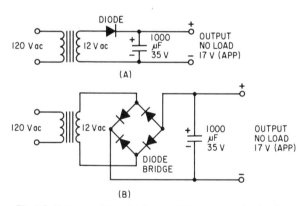

Fig. 5-3. Half-wave (A) and full-wave (B) power supply circuits.

sists of a transformer, a silicon diode, and a filter capacitor (see Figs. 5-3 and 5-4). The dc produced by such a supply tends to contain a considerable amount of ac ripple. Furthermore, this ripple percentage will increase markedly as the amount of power drawn from such a supply increases. In a like manner, the regulation of such a supply will tend to be poor, i.e., its output voltage with no-load will be high while the output voltage under full rated load will be low. In a half-wave capacitor filtered supply having a rating of 5 amperes, the output voltage with no-load may be as high as 17 volts, while the

Fig. 5-4. Typical electronic regulated power supply.

output voltage under full-load may sink as low as 10 volts. The problem of ripple and regulation is, of course, of minor importance when it comes to powering brute force devices such as bells, indicator lights, mechanical sirens, etc. With these devices, the only adverse effect would be the change in brightness or loudness as ripple and regulation parameters change.

The problem of ripple and regulation becomes of prime importance when the security control is to be used to power sophisticated space intrusion detectors. These devices very often have their own secondary regulation and ripple circuits but generally require an essentially clean dc input voltage. In addition, devices such as electronic sirens will not function properly or will sound rough and peculiar when the regulation and ripple of the power supply used to drive them exceeds required limits. Modern power supplies used in security controls employ a number of innovations to improve the regulation and keep the ripple, under full-load, at a minimum. Generally speaking, in simple supplies full-wave rectification with a large filter capacitor is preferred over half-wave rectification with small filter capacitors. In power supplies which are used to operate sophisticated loads, electronic methods must be employed to limit the ripple and give good regulation. Such electronically regulated supplies employ integrated circuits which contain the necessary amplifiers, voltage comparators, filtering circuits, Zener reference diodes necessary to achieve these parameters. Most of these integrated circuits are used for control purposes and are not capable of passing the large currents which are sometimes required. Consequently, these circuit chips are usually used in conjunction with heavily heat-sinked transistors to supply the required high currents.

Protecting Against Overloads

Regardless of whether the security control is ac operated, battery only, or a combination of both, the circuitry must be protected against overloads which can be caused by initial improper system wiring, shorts, or other failures. Refer to Table 5-3 for power source characteristics.

The most common way to protect the power supply system against such shorts is with a fuse. Such a fuse can either be of the standard type or of the slow-blow variety. The standard type fuse has a current rating and will blow when this current rating is exceeded. This fuse has the disadvantage that it might blow on momentary current demands which exceed its rating as a result of surges which are perfectly normal. For example, in a security system such a current surge may be encountered when a telephone dialer commences operation. This surge will last only for a second and a power supply can usually handle such momentary requirements. If an ordinary fuse is used in such a system it would blow rendering the system useless. A slow-blow fuse is a special type of fuse which tends to disregard momentary surges.

TABLE 5-3. POWER SOURCE CHARACTERISTICS

Parameter	Full-Wave Bridge With Capacitor Filtering	Electronic Regulation	Battery
Voltage output	Typically 3 - 24 V	Typically 3 - 24 V	Typically 6 - 24 V
Power output	Virtually unlimited	Virtually unlimited, becomes expensive above 10 A	Virtually unlimited
Load regulation	Poor	Excellent	Excellent
Line regulation	Poor	Excellent	Not applicable
Ripple under load	Poor	Excellent	Not applicable
Transient suppression	Poor	Excellent	Not applicable

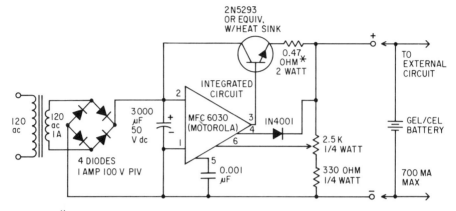

*CURRENT LIMIT SET TO APPROX. 700 MA BY 0.47 OHM RESISTOR. CURRENT LIMIT CAN BE CHANGED BY VARYING THIS RESISTOR. FOR USE AS A POWER SUPPLY, OUTPUT MUST BE 10 MV OR LESS WITH RATED LOAD OUTPUT CURRENT AND NO BATTERY CONNECTED ACROSS OUTPUT.

Fig. 5-5. Schematic of electronic regulated power supply.

Another method of protecting a power supply against overloads is through the use of a resettable circuit breaker. These devices have specific current ratings and generally operate on a thermal principle. Consequently, they have built-in protection against short-term overloads which they ignore. Circuit breakers, however, have the disadvantage of being somewhat sluggish and may ignore a slight overload for several seconds which sometimes is enough to damage the security system circuitry. Both fuse and circuit breakers have the additional disadvantage that, once they open, the system becomes inoperative until the fuse is replaced or the circuit breaker is reset. As a result, systems utilizing these protective devices should employ an audible or visible indicator to show that the protective device has blown and that the system is inoperative.

By far, the best method to protect power supplies from overloads is through the use of electronic current limiting circuitry (see Fig. 5-5). This type of circuit is utilized in the more sophisticated power supplies employing electronic regulation. The limiting circuitry continuously senses the amount of current demanded by the system. When an overload occurs, the current rises and a feedback signal is injected into the electronic regulation circuit. This signal immediately limits the total amount of current which the power supply will deliver even if the load becomes a dead short. This current limiting action remains effective until the short is removed and in no way damages the power supply circuitry. Thus, a typical electronically regulated power supply with current limiting may have a continuous rating of 4 amperes. If a short develops, the supply would hold the current to 4 amperes regardless of the load and with no damage to the circuitry. Normal operation will resume once the short is removed.

Automatic Standby Battery Systems

As we have previously discussed, a number of security controls do not utilize ac power supplies and operate entirely on batteries. With these controls, problems of ripple and regulation are, of course, not present. On the other hand, the batteries in such systems must be continuously watched to make sure that they remain fully charged. Furthermore, battery systems do not, of course, have a standby capability. Consequently, in most modern security systems ac power supplies are used. These power supplies are backed up by standby battery power which is automatically switched into the circuit in the event of power failure. This type of service uses either the ordinary dry cell or some form of rechargeable battery. In terms of the automatic switchover circuit in security controls using dry cell standby power, a diode, capable of handling the maximum current required, is inserted in series with the standby battery as shown. The diode keeps the ac supply power from flowing into the battery (Fig. 5-6A). When the ac power supply ceases to deliver power, current flows through the diode from the standby battery, automatically switching it into the circuit. This type of circuit, though very widely used, has the disadvantage in that the standby batteries can supply power to the system at the same time that the ac supply is supplying power. Such a condition will lead to the eventual partial discharge of the standby battery, and this situation may not be noted by the user unless the system is continuously tested for standby reserve. This problem occurs if the output of the ac power supply, because of poor regulation, drops below the battery voltage. If this condition occurs, the battery will deliver the difference between its open circuit voltage and the lower ac power supply voltage. Eventually the battery will stabilize at whatever voltage the power supply is generating, at which point the blocking diode once again enters into the circuit. Regrettably, in the event that the ac power supply comes back up to the proper voltage, the battery in such a circuit remains at the low value since no current can flow back into the battery through the diode. This essential drawback of the dry cell type standby power, in addition to the fact that dry cells have a limited shelf life, has led to the usage of rechargeable batteries which are continuously floated or trickle-charged by the ac power supply (Fig. 5-6C). Manufacturers of rechargeable batteries generally publish suggested circuits for keeping the batteries up to full charge and on the whole a float-charging technique is preferred over a trickle-charging circuit.

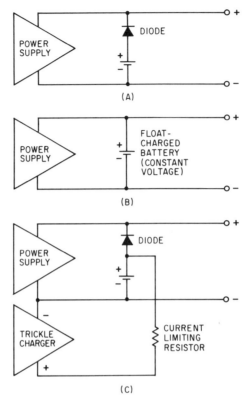

Fig. 5-6. Power supplies with automatic standby batteries.

Rechargeable Batteries

In terms of the types of batteries which are available, a number of choices exist. The most common battery is the standard carbon zinc battery (dry cell) which is available in a wide variety of voltages and current combinations. Rechargeable batteries are available in the traditional wet-cell type of storage battery (automobile batteries), Gel-Cells, and nickel cadmium. Each type of battery has its advantages and disadvantages and a number of factors must be taken into consideration in selecting power for these purposes. These factors include battery costs, capacity vs. size, and charge and discharge characteristics. One of the most important parameters with respect to battery power is the ampere-hour capacity of the battery in the event of power failure. All batteries have ampere-hour ratings. For example, a typical automobile wet cell storage battery of good quality might have a 60 ampere-hour rating. In general terms, ampere-hour means the amount of current a battery can deliver over a given period of time. Consequently, in the case of the automobile battery, a current of 1 ampere can be delivered by the battery for a period of 60 hours. This definition of ampere-hour ratings is somewhat simplified and should only serve as a general guide in attempting to determine the capabilities of batteries.

Battery Ampere-Hour Capacities

The actual capacity of a battery, in terms of stated ampere-hours, may not always be a simple multiple of the product of current and time. Deviations from the specified ampere-hour ratings do occur with the type of load that is involved, the timing cycles of the load, temperature, and other parameters. The data included in Figs. 5-7 and 5-8, show some typical capacities and resulting battery voltages which can be expected over time. As previously noted, the stated ampere-hour characteristics of a battery can be used as a general basic rule of thumb and all batteries, within reasonable limits,

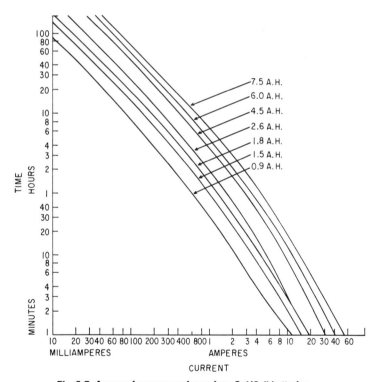

Fig. 5-7. Ampere-hour curves for various Gel/Cell batteries.

GC 610L-1 SPECIFICATIONS

1.	Nominal voltage	6 volts (3 cells in series)
2.	Nominal capacity at: 50 ma (20 hr. rate) to 5.25 volts 96 ma (10 hr. rate) to 5.13 volts 176 ma (5 hr. rate) to 5.07 volts 650 ma (1 hr. rate) to 4.80 volts	1.00 A.H. .96 A.H. .88 A.H. .65 A.H.
3.	Weight	.63 pounds (approximate)
4.	Energy density (20 hr. rate)	.86 Watt-Hours/Cubic Inch
5.	Specific energy (20 hr. rate)	9.7 Watt-Hours/Pound
6.	Internal resistance of charged battery	Approximately 90 Milliohms
7.	Maximum discharge current with standard terminals	40 amperes
8.	Operating temperature range: Discharge Charge	 −76°F to +140°F − 4°F to +122°F
9.	Charge retention (shelf life) at 68° F 1 month 3 months 6 months	 97% 91% 82%
10.	Sealed construction — can be operated, charged or stored in ANY position without leakage of any corrosive liquid or gas. Battery protected against internal pressure build-up by self-sealing vents which pass only dry gas.	
11.	Terminal — solder lug or use as "slip-on" tab. Will accept AMP, Inc. Faston "187" Series (for .032" tab) receptacles or equivalent.	
12.	Case material — High impact polystyrene, light gray in color.	

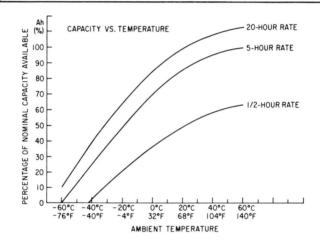

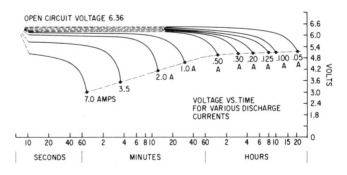

Fig. 5-8. Typical discharge curves for Gel/Cell batteries.

have the ability to deliver much higher power for shorter periods of time. Thus, in the case of the car battery, it could deliver 2 amperes for 30 hours (24 watts) or 4 amperes for 15 hours (60 watts) or, to cite an extreme case, as much as 500 amperes for 5 minutes (6,000 watts). Consequently, in computing the amount of time for which a standby power source will function, it is necessary to determine the system wattage load in the continuous or in the alarm state. Table 5-4 shows typical ampere-hour capabilities of a variety of batteries.

TABLE 5-4. TYPICAL AMPERE HOUR CAPACITIES FOR VARIOUS BATTERIES

Type	Ampere-Hours
Average automobile battery	60
Globe model GC660-1, 6 V rechargeable Gel-Cell battery	6
Gould model PP2312, 12 V rechargeable nickel cadmium battery	23
Eagle-Picher model CF6V2.5, 6 V rechargeable lead dioxide battery	2.5
Gould Gelyte model PP690, 6 V rechargeable battery	9
Two (Eveready 510 or equal) 6 V lantern dry cells in series	3
Two (Eveready 1461 or equal) 6 V hot shot dry cells in series	10

All batteries do have a limited shelf life, which means that their voltage drops after a period of time, even when they have not been used. Standard dry cells generally have the shortest shelf life and should be replaced, conservatively speaking, every 6 months. Rechargeable batteries also have a certain shelf life but this tends to be very long, often in excess of years. Furthermore, the rechargeable batteries of the Gel-Cell and nickel cadmium type cannot be recharged an infinite number of times. The recharging cycle of such batteries is usually limited to 200 recharging cycles before the battery will have to be replaced. Even this number is dependent upon the manner in which the battery is charged and discharged. Frequent, rapid, high-current charging of a completely dead battery tends to shorten the life of the battery and the number of recharge cycles which one can reasonably expect. Float and trickle charging at the suggested manufacturer's rate over long time periods (12 hours or more) tend to lengthen the life of the battery. Rechargeable batteries, such as the Gel-Cells and nickel cadmium type can, of course, be operated in any position and do not need special venting. The traditional rechargeable wet cell or storage battery, such as is used in automobiles, has the highest volt-ampere capacity at the most reasonable cost. However, these batteries can only be operated in the horizontal position and their liquid electrolyte must be periodically checked.

6
Local Alarms

After an intrusion has been sensed and processed through the control circuitry, an alarm condition must result in the security system. The function of this alarm condition is two-fold since intrusion indication must be given on a local and remote basis. The local alarm indication is audible and/or visible, warning the intruder that he has been detected and possibly deterring him from doing any further damage. Furthermore, the local alarm must also alert nearby patrolmen, guards, passersby, neighbors, and others who might apprehend and deter the unauthorized person. In addition, the system must also transmit an alarm signal, usually via telephone lines, to some remote strategic point, such as a police station or central station. Remote alarm indicating equipment will be discussed in Chapter 7; this chapter concerns itself only with the local alarm.

A number of basic local alarms are currently used and these include:
1. Bells
2. Electronic sirens
3. Mechanical sirens
4. Resonating horns
5. Beacons

Parameters

In selecting the proper local alarm, a number of parameters must be taken into consideration. It is the function of the local alarm to frighten the intruder and to alert others; consequently, such a device should be as loud and as distinctive as possible. The level of intensity which is required is, of course, dependent upon the area which must be covered and the associated noise levels which exist. It is generally true, especially where large areas are to be covered, that multiple local alarms of lower intensity will give better coverage than a single local alarm of high intensity.

In outdoor service requiring coverage beyond half a mile, large sirens or air horns are useful. The smaller sirens and air horns can be used in the range of 5,000 feet down to 1,000 feet. Below that range electronic sirens, bells, and beacons are useful. With respect to indoor local alarms, large open spaces can be treated the same way as outdoor areas. When coverage is required in partitioned spaces, a number of individual low power local alarms are usually employed. In terms of loudness, regardless of the kind of local alarm which is used, a number of characteristics must be taken into consideration.

If the local alarm is electrically operated, there is a direct relationship between the input wattage which the device consumes and the amount of output power which it radiates. It is self-evident that some devices are more efficient than others in converting input power to output power, but on the whole, overall efficiencies are generally of the same order of magnitude. Table 6-1 shows some typical voltage, current, wattage consumption figures which can be expected in electrically operated local alarms. Note that data has only been given for 6-volt and 12-volt devices. Local alarms are, of course, available, operating at other voltages including 120 volts ac. However, 120 volts ac operated local alarms are not recommended since their installation requires the strict adherence to local electrical codes and requires installation by a licensed electrician. In analyzing the figures given in the Table 6-1, it must be pointed out that regardless of device efficiencies and other parameters which may come into play, it is unlikely that a typical electronic siren drawing 1 ampere at 12 volts (12 watts) will be louder than a motor driven siren which draws 10 amperes at 12 volts (120 watts)—regardless of what the manufacturer might state with respect to loudness.

TABLE 6-1.
TYPICAL VOLTAGE, CURRENT, AND POWER CONSUMED BY LOCAL ALARMS

Typical Device	DC Voltage	Current	Watts
6" bell	6	700 mA	4.2
6" bell	12	300 mA	3.6
10" bell	12	400 mA	4.8
Motor-driven siren	12	10 A	120.
Motor-driven siren	6	20 A	120.
Electronic siren	12	1 A	12.
Electronic siren	6	2 A	12.
Beacon	12	3A	66.

TABLE 6-2. LOUDNESS OF LOCAL ALARMS

Device	Typical Loudness in Decibels (Measured at 10 ft)
6" bell	100
10" bell	102
Motor driven siren	120
Resonating horn	105
Compressed air horn	127
Electronic siren	101

The loudness of a local alarm may not always be stated in terms of input power, but sometimes is stated in decibels at 10 feet, as Table 6-2 shows. For these measurements, a microphone is placed in front of the device in free space and the sound intensity is measured at that point. (For comparison, Table 6-3 gives an overall picture, in terms of dB, of the loudness of various

TABLE 6-3. TYPICAL SOUND LEVELS

Noise Source (Distance From It)	Decibels
Studio for sound pictures	20
Studio—speech	30
Soft whisper (5 ft)	32
Minimum levels—residential areas in Chicago at night	40
Private business office	50
Light traffic (100 ft)	50
Average residence	50
Large transformer (200 ft)	50
Near freeway—auto traffic	60
Large store	60
Accounting office	60
Freight train (100 ft)	70
Vacuum cleaner (10 ft)	70
Speech (1 ft)	70
Tabulating room	80
Inside sports car—50 mph	80
Pneumatic drill (50 ft)	82
Boiler room	90
Printing press plant	90
Textile weaving plant	90
Subway train (20 ft)	90
Electric furnace area	100
Cut-off saw	105
Pneumatic hammer	105
Casting shakeout area	110
Riveting machine	110
Jet takeoff (200 ft)	122
50-hp siren (100 ft)	132

familiar sounds.) Local alarm loudness data, stated in db, can also be misleading. For example, the actual loudness of a local alarm is not only dependent upon the dB rating, but also on the frequency, wave shape, and duty cycle of the signal involved. The microphone which is used in such dB measurements is linear with respect to frequency. However, the human ear is not, and becomes more sensitive around 1,000 hertz. Figure 6-1 shows the loudness sensation which the listener experiences as the frequency broadens in a constant amplitude signal. To human ears, broad band sounds like those of jet engines seem much louder than pure single frequency sounds. Furthermore, broad band sounds, i.e., sounds with high harmonic content, seem louder if their spectral content favors the higher frequencies above 160 hertz. Consequently, a broad band signal between 160 hertz and 670 hertz will not seem as loud as a broad band signal between 670 hertz and 2,000 hertz. Let us take another example, the case of two sirens, one having a center frequency of 600 hertz, while the other has a center frequency of 1,000 hertz. It is quite possible, in a traditional dB-microphone test, that these two sirens will have the same loudness rating. However, when subjected to listening tests the siren having the 1,000 hertz center frequency will sound louder than the siren having the 600Hz center frequency because of the basic nonlinearity of the human ear. Although commercial sirens are available with a wide variety of frequencies, most of them have their center frequency in the 1,000Hz region. Another factor which does not show up in the dB rating is the effect of the local alarm's wave shape. Here again, the human ear tends to be irritated more by frequencies which are rich in higher order harmonics. Consequently, an electronic siren operating at 1,000Hz and having a pure sine wave output will tend to be less irritating than a siren operating at the same frequency having a sawtooth (distorted) output. This is due to the fact that a sine wave is essentially harmonic free while sawtooth waves are rich in harmonics. Furthermore, the device's duty cycle also affects its loudness.

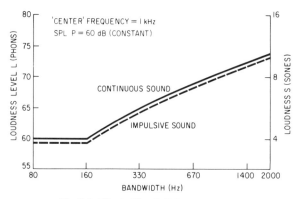

Fig. 6-1. Effect of bandwidth on loudness.

For example, the dB measuring technique is essentially an integrating process where the loudness of the device is averaged out over time. Let us take the case of two resonating horns. The first emits high intensity bursts of short duration with long off-times (refer to Fig. 6-2). The second emits medium intensity bursts but at a more rapid rate. To the ear, the high intensity long off-time device will sound louder than the rapidly pulsing medium intensity device. However, such listener-generated information will not be entirely confirmed by microphone-dB developed data since the measuring system may not be able to follow the brief high intensity bursts and, consequently, tends to average them out thereby giving the high intensity horn a poorer rating when it is in fact louder.

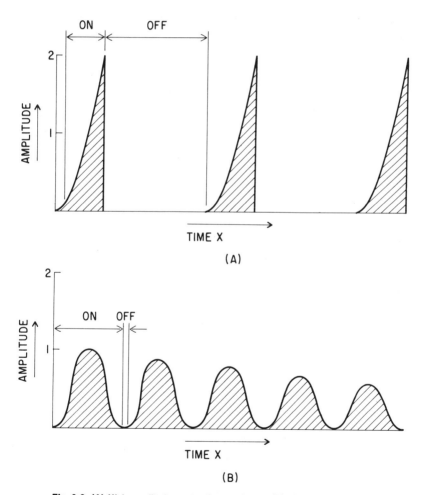

Fig. 6-2. (A) High-amplitude sawtooth wave forms. (B) sine wave forms.

Another factor to consider is the efficiency with which the device is coupled to the air. This is determined by the physical configuration of the speaker or horn which is used. In the case of sirens, long straight horns with gradual flares and small mouth apertures are moderately directional. Devices using short horns, such as the traditional loudspeaker trumpets, with mouth apertures of 12 inches, tend to be omnidirectional. The transfer efficiency of such alarms can be improved by designing a horn specifically around the operating frequency of the device. Furthermore, omnidirectional effects can be achieved in basically directional devices by using a multiplicity of horns ranged around a central driving mechanism in a circular fashion. Lastly, it must be remembered that all published data is usually developed in still air. The actual loudness of any device is also dependent upon temperature and moisture content of the air as well as wind direction, and these factors also should be considered in selecting a proper local alarm.

In summary, the loudness or intensity of a local alarm can, to some degree, be evaluated by wattage and dB data as published by the manufacturer. However, as Table 6-2 indicates, this data may be very close, and in some instances, may be misleading. In the final analysis, it is best practice to perform actual listening tests on location in order to select the local alarm which best fulfills the requirements.

Alarm Bells

Of all the local alarm devices used, bells are by far the most popular. Such bells usually come from 4 inches to 10 inches in diameter, the size of the bell used being dependent upon the loudness which is required. Bells normally employ the traditional solenoid interruption mechanism as shown in Fig. 6-3 and generally operate from 6 or 12 volts dc. With respect to current requirements, these bells consume anywhere from 250 MA to as much as 1 ampere. As is the case with other local alarm devices, bells are also available for 120 volt ac operation. However, 120 volt ac bells are not widely used since their installation usually requires the services of an electrician who has to wire in accordance with the local electrical code.

Bells used for security systems are generally of the underdome type which has the entire electrical clapper mechanism located inside of the gong (Fig. 6-4). This type of bell gives the maximum amount of protection against the adverse effects of weather,, and some types actually can be used outdoors without any other protective enclosure. However, for the majority of security bells, it is suggested that a suitable weatherproof box with louvers be employed when the bell is mounted outside. As can be seen from the basic bell diagram in Fig. 6-3, this type of circuit switches a solenoid in and out at a rapid rate, causing the bell to ring. The rapid insertion and removal of such an inductive load will generate pronounced spikes on the power supply line

LOCAL ALARMS 105

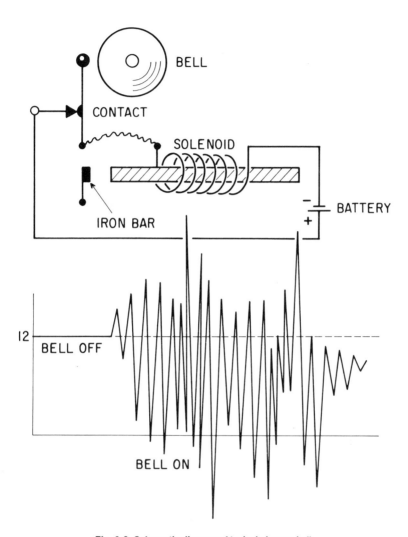

Fig. 6-3. Schematic diagram of typical clapper-bell.

and these spikes may give trouble in the control circuitry, especially if the control utilizes transistors and silicon controlled rectifiers (SCR). The adverse net effect of a bell in such a circuit may be that the bell permanently latches the control into the alarm state once the bell starts ringing, thereby defeating any automatic recycling features. The electrical spike and noise induced by bells, and their effect upon transistorized circuitry, is very difficult to predict, so that a trial and error technique in conjunction with good bypassing practice is advised. Therefore, it is usually advisable to furnish a separate power-supply source, such as batteries, for powering bells.

Fig. 6-4. Alarm bell of underdome construction showing solenoid clapper mechanism.

Electronic Sirens

Next to the bell, the most popular local alarm is the electronic siren. In this device, some form of transistorized oscillator circuit is used in conjunction with a power amplifier that connects to a specially designed loudspeaker (Fig. 6-5) which emits the siren tone. The advantage of this type of siren lies in the fact that it is extremely reliable, quite weatherproof, and draws comparatively little current. Sirens of this type usually operate from 6 to 12 volts and generally do not draw more than 1.5 amperes, a majority of them requiring about .750 ampere.

A typical electronic siren circuit is shown in Fig. 6-6. This circuit consists of a simple multivibrator type oscillator which generates the center frequency. Practice has indicated that the most effective siren frequencies are from 600 to 1,000 hertz. The output of this oscillator is fed to a voltage-sensitive amplifier which determines the warble or frequency glide rate of the siren. The signal is then fed to a Darlington type power amplifier whose output connects to a suitable loudspeaker, to serve as the siren.

The total power consumed by the aforementioned circuit depends upon the impedance of the loudspeaker's voice coil. For instance, in Fig. 6-6, let us assume that the voice coil (A) has an impedance of 8 ohms at 1,000 hertz. From Table 6-1, it is seen that an electronic siren operating from a 12 volt dc source will draw 1 ampere. This current will flow through the voice coil and

LOCAL ALARMS 107

Fig. 6-5. Federal #300 electronic siren (electronic equipment is in lower case).

the 10-ohm resistor in series with the 12-volt supply. From Ohm's Law, R = E/I, the total dc resistance of the voice coil is (12−10)=2 ohms. The effective ac voltage, at 1,000 Hz, developed across the voice coil, in this particular case, is computed as about 4 volts. Thus, the power drawn by the loudspeaker will be:

$$P = \frac{E^2}{Z} = \frac{4^2}{8} = 2 \text{ watts}$$

To provide more power for a louder siren output, it is desirable to lower the impedance of the voice coil. This is not practical, however, because of the difficulty in constructing a loudspeaker with such a low-impedance voice coil. The effective method is to increase the ac voltage across the loudspeaker's voice coil by the use of a step-up output transformer. This type of transformer should have a primary impedance (Zp) of 2 ohms and a secondary impedance (Zs) of 16 ohms. The resultant turns ratio and ac voltage increase, from the usual transformer formula are:

$$Ns = \sqrt{\frac{Zs}{Zp}} = \sqrt{\frac{16}{2}} = 2.8 \text{ times}$$

where: Ns is the number of turns in the secondary of the transformer as compared to the primary,

108 ELECTRONIC SECURITY SYSTEMS

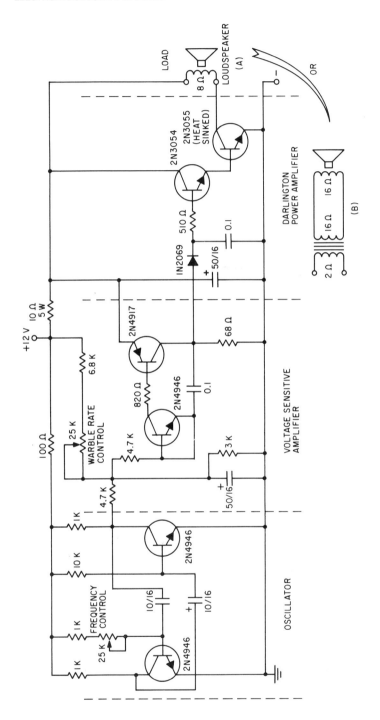

Fig. 6-6. Schematic diagram of electronic siren.

In addition, it will be necessary to reduce the value of the resistor in series with the 12 volt supply from 10 ohms to about 7.5 ohms. This will permit more current to flow through the primary of the output transformer in order to provide additional power.

With the above changes, the direct current through the primary of the output transformer, per Fig. 6-6B, will be 1.5 amperes. The ac (1,000Hz) voltage developed across the 16 ohm voice coil impedance will be 17 volts. Consequently, the power taken by the loudspeaker in this revised circuit will be, from the formula for power:

$$P = E^2/Z = (17)^2/16 = 18 \text{ watts}$$

It is assumed, in this connection, that the system's power supply is capable of delivering the additional required power.

Still another advantage of electronic sirens lies in the fact that a wide variety of tone colorations can be produced in the oscillator portion of the circuit. Various oscillator designs offer sine waves, sawtooth waves, square waves, complex pulses, all at various frequencies. It consequently becomes possible, and a number of sirens employ this technique, to furnish the oscilator as a plug-in module, giving the user a wide variety of siren sounds, including wails, yelps, intermittent beeps, stutter, horn sounds, and many more. Furthermore, many of these circuits have adjustments in which the user can select the siren's center operating frequency and warble rate to suit the application requirements.

Mechanical Sirens

By far the loudest local alarm can be developed through the usage of a mechanical siren (see Fig. 6-7). Mechanical sirens of this type may use a single horn or a series of horns for the purpose of spreading the signal, all driven by the same source. The siren sound generated by this device can be developed by a number of means. One of the most popular methods is the usage of a slotted rotating wheel (or squirrel cage wheel) driven by a high speed motor. The wheel chops up the air, developing the siren tone. The frequency of the tone, in other words the warble rate, is developed by increasing and decreasing the speed of the motor. Another version of this type of siren uses compressed air to rotate the sound producing wheel and the compressor may be located next to the siren or at some remote point. In still another type, compressed air is forced through a tuned cavity, thereby generating a siren sound without the usage of a rotating wheel. (This technique is similar to the sound generating mechanism used in organ pipes except that the frequency is, of course, much higher.) The compressed air which is required by this type of

110 ELECTRONIC SECURITY SYSTEMS

Fig. 6-7. Typical mechanical or motor-driven siren.

siren may be generated by a centrally located compressor and storage tank or may be delivered by a small compressor mounted as part of the siren.

Furthermore, compressed air sirens are currently available in which the air is stored in a replaceable canister (aerosol can) right beneath the siren. An electrically operated solenoid mechanism is used to trigger the siren which will operate until the canister is empty. This type of siren has a number of advantages and among them is the fact that, from an electrical point of view, it draws very little power—only enough to trip the solenoid. In addition, with this kind of siren, a tamper loop can be wired into the solenoid control cable. Wiring of this tamper loop can be such that the siren will be energized whenever the cable is cut. (This is obviously not possible with other sirens that get their operating power from a remote point.) This type of siren has the disadvantage that it will only run while the air supply from the canister lasts. Once the siren has triggered and run, the canister must be replaced, which may be a maintenance problem.

In summary, mechanical sirens do offer the loudest form of local alarm. With the exception of the aerosol type siren, mechanical sirens consume a considerable amount of power, with the smallest usually drawing no less than 120 watts, while the very largest may require thousands of watts to operate.

LOCAL ALARMS 111

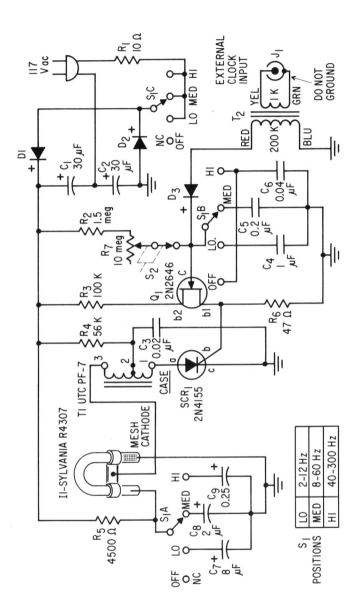

Fig. 6-8. Schematic diagram of Xenon Beacon Flasher Unit.

Other Local Alarms

The resonating horn local alarm emits a sound similar to the typical automobile and truck horn. Resonating horns have a diaphragm which vibrates back and forth emitting a horn sound. This sound is then coupled to a trumpet projector which radiates into the air. The diaphragm in low and medium power resonating horns is often electrically driven by a solenoid mechanism. In high intensity resonating horns the diaphragm may be driven by a motor. (Resonating horns are also available in which air driven resonating cavities are used.)

A rotating beacon or some other form of omnidirectional, flashing, high intensity light source is also often used as a local alarm indicator. Such beacons are mounted outdoors as high as possible so that they will have the best chance of attracting attention. The beacons may consist of a high intensity rotating incandescent light in a weatherproof enclosure, or a nonrotating mercury vapor or xenon flash tube. The effectiveness of these devices is again a function of their intensity which is proportional to the input wattage which they consume. The intensity is often stated in candle power which, for a typical 12-volt beacon, may range from 10,000 to over 50,000. Beacons of this type are usually available with a choice of dome colors. Furthermore, in the case of the omnidirectional flash tube beacons, an adjustment is usually available which permits the user to adjust the flash rate from roughly 50 to 100 flashes per minute (Fig. 6-8).

7
Remote Alarms

In the previous chapter we saw that the local alarm, the bell, or siren frightens away the intruder and warns nearby personnel that an illegal entry has been made. Virtually every security system, in addition to the local alarm, must have a remote alarm signal which is also generated when the premises are violated. This remote alarm signal is transmitted to a distant point where it is monitored and from where corrective action is taken. Such remote monitoring points may be police stations, privately-owned central stations, guard services, or even the home of key personnel and friends.

Transmission of the remote alarm signal is almost always accomplished by means of commercial telephone lines and with such a system, two types of lines are used: (1) a customer's telephone line, and (2) a private or direct leased telephone line.

A number of devices are available which send the remote alarm signal over telephone lines. All of them are triggered into action by the security control which invariably contains a set of contacts for the sole purpose of energizing the remote alarm system. Each remote alarm device will be discussed in detail. Table 7-1 has been included to show the general characteristics of the devices and the type of telephone line with which they are usually associated.

By far the most commonly used remote alarm device is the telephone dialer which operates with the customer's telephone lines. Such dialers can be subdivided into two groups: (1) those which work with a tape cartridge, and (2) dialers which dial and transmit information through digital pulsing.

TABLE 7-1.
GENERAL CHARACTERISTICS OF REMOTE ALARM DEVICES

Type of Device	Telephone Line	Transmitted Data	Coupler Needed
Tape telephone dialer	Customer	Voice	Yes
Digital dialer	Customer	Coded (ac)	Yes
Reversing system	Leased	Coded (dc)	No
McCulloh loop	Leased	Coded (dc)	No

Tape Telephone Dialer

The tape-type dialer is in effect nothing else but a special tape player utilizing an endless tape cartridge or a standard tape cassette (Fig. 7-1). These dialers contain a motor operated tape drive mechanism and the necessary electronics to amplify and transmit the signals, recorded on the tape, over telephone lines. Furthermore, these dialers contain control circuitry which makes it possible to start, stop, test, and automatically recycle the device. The actual alarm message of such a dialer is in the form of a prerecorded magnetic tape. Some of the more elaborate dialers have their own programming-recording system built into the dialer (Fig. 7-2). However, a majority of the dialers currently in use have their tapes programmed through a programmer which is usually available as an accessory.

In terms of tape programming for this type of telephone dialer, the cartridges which are employed offer a wide variety of message lengths and number calling capabilities. Every tape cartridge must, of course, have

Fig. 7-1. Record-O-Phone #ADS 800 dialer, programmer, and audio monitoring system.

REMOTE ALARMS 115

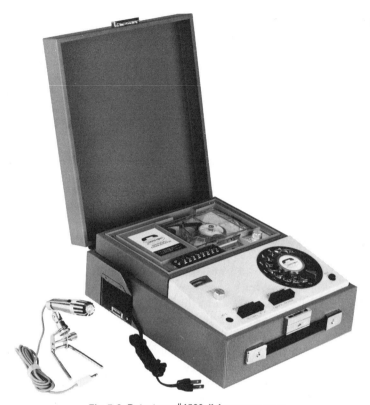

Fig. 7-2. Detectron #1520 dialer programmer.

recorded on it the preprogrammed number which the dialer must dial in the event an intrusion occurs. This data is entered on the tape through the programmer and may be in the form of digital pulses or tone sequences as shown. These pulses or tone sequences are generated in an oscillator circuit in the programmer and are recorded on the tape through the use of a standard telephone dial which is used for programming. When the tape cartridge starts in such a dialer, it initially dials out the prerecorded number. At the remote point, the telephone begins to ring and the dialer cartridge is arranged to run for a sufficiently long enough time for a normal ringing sequence to occur. Once the time for the ringing sequence has elapsed, a message announcing the intrusion is recorded on the tape. The time for the ringing sequence is arbitrary. Typical tape cartridges for these dialers are usually programmed to leave time for four to six rings at the remote location. Obviously, if, at the remote location, someone answers the telephone after the first ring in a system which has been programmed for four rings, he will not hear the message until the time allocated for the remaining three rings has elapsed. Once this time has elapsed, the remote point will hear the message and, consequently, can take corrective action. In general, this message is repeated three or four times on the tape before the telephone line release signal is inserted into the tape cartridge. The purpose is to make certain that the message is understood.

The telephone release signal is given in the form of a pulse or a tone and it electronically releases the telephone line at the place of intrusion and rearms the telephone dialer. A wide variety of programs can, of course, be generated utilizing such tape cartridges. Programs in this type of dialer need not be limited to dialing only one number and citing only one type of emergency, such as burglary. For example, cartridge-type telephone dialers are available using two-track tape for which the first track may be used for the announcement of night-time burglaries while the second track may be used for the announcement of daytime panic situations. Furthermore, tape cartridges are available having running times from as little as 30 seconds to as much as 10 minutes. Obviously, on the longer tape cartridges, a telephone number dialing sequence can be repeated at 2-minute intervals for the entire 10-minute period, making certain that the message gets through in the event that the number called is initially busy.

The dialer itself is energized by the alarm signal emanating from the security control. In most telephone dialers used for security purposes, the energizing signal may be a simple relay closure or the application of voltage. The dialer's power is generally obtained through the device's own internal power supply, and here again standby battery systems may be included so that the dialer will operate even in the event of power failure. Another feature, which is usually included in the better type of cartridge telephone dialers, is the line seizure capability. This is a special circuit which permits the dialer to send out its signal even if the intruder is clever enough to lift the telephone receiver off-hook immediately upon entering the premises.

The schematic for a typical tape-type dialer is shown in Fig. 7-3. Note that this is a two-channel dialer working from standard ¼ inch magnetic tape, upper channel A, and lower channel B. The signals from either channel are amplified by transistors Q1 and Q2 in order to pulse a dialing relay. The dialing signal on the tape for this type of dialer is a brief burst of 1,000 Hz tone which triggers the pulsing relay. The dialer has its own internal full-wave power supply but does have capability for automatically switching in a 12-volt standby battery of the dry cell type, which is isolated from the circuit through diode CR3. The control switch on the dialer has three positions—OFF, TEST, and ON. In the TEST position, the dialer is immediately energized but the telephone output circuit is disabled. The signal on the tape can then be monitored through the built-in monitor speaker. In this particular dialer, automatic shut off and recycling is accomplished by means of a piece of conductive foil tape in the tape cartridge. When this tape passes across feel fingers SW2, a circuit is completed causing the dialer to halt and rearm itself.

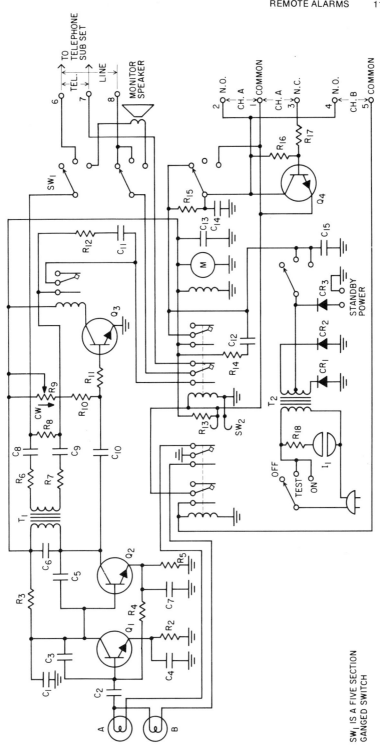

Fig. 7-3. Schematic of two-channel telephone dialer unit.

In the ON position the monitor circuit with its test speaker is disabled and the telephone pulsing and audio circuits are energized. Channel A of this particular dialer can be energized by shorting Point 1 to Point 2 or opening Point 1 to Point 3. The B channel is energized by shorting Point 5 to Point 4. The circuit is arranged in such a manner that if both channels are energized simultaneously, channel B will take preference over channel A and, consequently, that message should be one having the higher priority. In order to achieve the proper impedance and signal amplitude required by the telephone company, an output transformer, T1, and resistors R6 and R7 are employed in the circuit which is connected to the telephone line. This particular dialer is designed so that it can actually be wired directly across the telephone line without the need of any telephone company interface equipment. Virtually all telephone dialers can be used that way; however, it is generally the telephone company's position that a coupler be used as an interface device between the dialer and the telephone line. This coupler is installed by the telephone company and leased to the customer at a monthly rate. The particular coupler which is discussed later, usually requires 18 volts dc for its operation. This voltage must be supplied by the security system.

Digital Telephone Dialer

The purpose of a digital telephone dialer is exactly the same as that of a tape cartridge telephone dialer except that the dialing sequence and the subsequent message is generated entirely through electronic digital circuitry without any moving parts. The advantage of a digital dialer primarily lies in the fact that it is generally more reliable than its tape counterpart. Furthermore, it has the programming circuits built directly into the unit through a series of switches, so that the user can program the dialer by simply setting the switches, without having to have a separate tape programmer. Thus, digital dialers generate the necessary dialing sequence pulses entirely through electronic circuitry, without the use of magnetic tape (Figs. 7-4 and 7-5).

Since a digital technique is used, this type of dialer cannot give a voiced message. Information with respect to the location of the place of intrusion, time of intrusion, nature of intrusion, etc., however, may be developed in the digital dialer by means of a tone code. Digital telephone dialers generate their various coded signals entirely by means of integrated circuits. These circuits contain a basic clock frequency which contains an oscillator running at a specific fixed frequency. A series of staggered, frequency dividing IC's are then used to generate the dialing pulses and the information pulses which are required.

The block diagram (Fig. 7-6) shows a typical digital dialer. This particular dialer is activated by shorting the input trigger circuit which in turn latches an SCR in series with the battery. This particular dialer is entirely

REMOTE ALARMS 119

Fig. 7-4. Automatic digital dialer device.

Fig. 7-5. Remote station digital receiver decoder and number display unit.

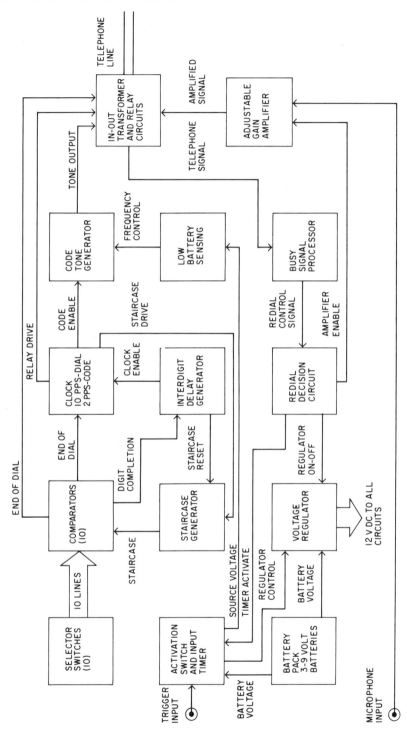

Fig. 7-6. Block diagram of digital telephone dialer.

battery operated and, consequently, runs totally independent of line-voltage fluctuations or failures. The amount of power drawn by this device, when it is not dialing, is so low that shelf life of between 6 months and 1 year can be expected from the batteries used. The dialer employs two 9-volt batteries connected in series so that 18 volts are developed. This voltage is fed through a voltage regulator which delivers 12 volts to the circuitry. (In such a circuit a margin of safety exists even in the event that the battery voltage drops.) Once the SCR has fired, the output voltage is fed through the regulator which in turn supplies the remaining circuitry. Upon receipt of this operating voltage, the dialer output relay closes and delivers an off-hook signal to the telephone line to which it is connected, directly or through a coupler.

The clock oscillator circuit in this particular digital dialer pulses the output relay at ten pulses per second. This pulsing rate does not commence until a brief time delay has elapsed to allow for the dial tone to be received from the central office. Once the clock oscillator starts pulsing the relay, a staircase generator begins to develop a staircase voltage whose steps and amplitude are controlled by the clock frequency. This staircase voltage is fed to all of the ten integrated-circuit comparators, but the circuitry is such that only the first comparator is active. Each of the ten comparators has an external selector switch which can be set to any number from 1 through 10 and off. (Fig. 7-7 shows a code chart for digital dialers.) When the staircase volt-

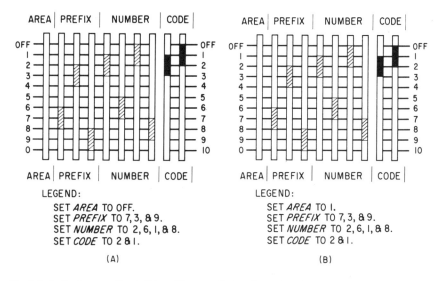

Fig. 7-7. Code chart for digital dialer, (A) Local dialing. *Example:* 739— 2618. (B) Distance dialing with common area code. *Example:* 1— 739— 2618.

age reaches the comparator switching level, as set by the comparator's selector switch, the comparator fires and latches and then sends a signal to the interdigit delay. The interdigit delay causes the clock oscillator to stop, and also automatically resets the staircase generator.

Upon completion of this cycle, the clock oscillator starts again and the second comparator becomes active once again to trigger when the staircase generator reaches the second comparator's triggering level as set by the selector switch. Once this condition is attained, recycling takes place and the sequence commences for the third comparator. This entire process sweeps through the eight comparators which, in this particular digital dialer, have been set aside for dialing sequences. Once the dialing sequence has been sent out by this system, the circuit switches over and allows the identifying code signal to be fed into the telephone line. In this particular digital dialer, two sequential tone codes are available.

The frequency of these two tone codes can be varied from approximately 300 to 3,000 hertz and the desired frequencies can be set with the switches associated with the ninth and tenth comparators. These comparators work the same way as the number-dialing comparators utilizing the clock and the staircase generator. However, for purposes of easy code identification, when the last two comparators are operative, the clock rate is automatically reduced from ten pulses per second to two pulses per second. This means that the tone bursts which are generated by the ninth and tenth comparators are in the audio-frequency range and can be used to trigger identifying circuits, such as frequency-sensitive relays at the remote point. This particular digital dialer has some unique additional features which are highly desirable. One of these is the fact that the identifying code signals will repeat their message continuously. This means that if the identifying code is not immediately recognized at the remote point, the dialer repeats the code over and over again. Furthermore, the circuitry is also designed in a way that an accessory microphone can be used instead of the identifying code. When the code selector switches on the ninth and tenth comparators, they are set to the OFF position and a high-gain audio amplifier is automatically switched into the circuit. This amplifier has a microphone input jack to which any sensitive high impedance microphone may be connected. When the dialer is operated in this mode, the microphone will pick up and send to the telephone lines anything it hears. Consequently, the system becomes a kind of audio-detection sensor in which the comparator at the remote point can actually listen to what is going on in the protected premises. In addition, the dialer has circuitry which automatically recycles it in the event that the number initially dialed is busy. After the initial dialing sequence is over and before the code signal is injected, a busy-signal recognition circuit senses if the number being called is ringing or busy. If busy, the circuit automatically reduces the regulator voltage to zero which resets and rearms all circuits. Providing that the input trigger is still shorted, the dialer will redial over and over again until the number being called is no longer busy.

With respect to recognizing the alarm signal at the remote location, a number of techniques are possible. The simplest technique is, of course, the one employed with the tape-type dialer in which a voiced message is delivered via the normal telephone. This voice message must, of course, be understood by persons listening (policemen, answering services, detective agencies, friends, etc.) who then take the required corrective action. In the vast majority of cases, voice messages are employed with tape-type dialers. It is, of course, possible to record tone-encoded messages on a tape-type dialer, but this is generally not done.

In tone-encoded messages, such as must be used in digital dialers, remote-point identification can also be done by human listening. If human listening is used, such a process has obvious dangers and limits. First of all, the listener must have a pretty good sense of pitch to identify the particular premises-code involved. If the remote point is used to monitor only one dialer, tone encoding is feasible. For instance, a telephone answering service, into which a tone-encoded dialer is programmed, would know that if the telephone rang and a 1,000Hz audio tone was repeated, an intrusion had taken place. Obviously, this system becomes completely unrealistic when a number of locations are being monitored. In most applications, the remote monitoring point is an answering service, a central station, or a police station which literally may monitor thousands of dialers.

Tape-type voice messages are, of course, possible and are used in such remote locations. Voice messages have the disadvantage that they take a fairly long amount of time to deliver and depend upon human hearing for recognition. In addition, if records are to be kept, it then becomes necessary with voice messages for the listener at the remote point to enter the message in some form of log. This type of programming is cumbersome, expensive, and open to human error. Consequently, where a large number of dialers are involved, tone-encoded messages are usually preferred. As previously indicated, with tone-encoded messages the signal is received at the remote station by some electronic sensing circuit which is frequency sensitive. The number of tone-encoded premises which can be identified is virtually unlimited and depends primarily upon the complexity of the tone code. (Tone codes are given in Table 7-2.) In general, if a single tone is used, about thirty different premises can be identified since the tone-generating circuits usually deliver signals from 300 to 2,500Hz. In such a system, each premise is assigned a specific frequency starting with 300 and ending with 2,500Hz with the unused frequencies in between acting as guard bands. If two tone-encoded signals are used, the system capability becomes much larger (30 x 30 = 900) and if three tone-encoded signals are used (30 x 30 x 30) the system has a capability of 27,000 and so forth, in accordance with the equation 10^n.

ELECTRONIC SECURITY SYSTEMS

TABLE 7-2. TONE CODES

Freq. Hz	Freq. Hz	Freq. Hz
67.0	330.5	1084.0
71.9	339.6	1120.0
77.0	349.0	1151.0
82.5	358.6	1190.0
88.5	368.5	1220.0
94.8	378.6	1265.0
100.0	389.0	1281.0
103.5	399.8	1291.4
107.2	410.8	1320.0
110.9	422.1	1355.0
114.9	433.7	1400.0
118.8	445.7	1425.0
123.0	470.5	1430.5
127.3	483.5	1450.0
131.8	496.8	1500.0
136.5	510.5	1520.0
141.3	524.6	1550.0
146.2	539.0	1585.0
151.4	553.9	1600.0
156.7	569.1	1650.0
162.2	582.1	1750.0
167.9	600.9	1764.0
173.8	617.4	1850.0
179.9	634.5	1963.0
186.2	651.9	1950.0
192.8	669.9	2050.0
203.5	688.3	2150.0
210.7	707.3	2175.0
218.1	726.8	2185.0
225.7	746.8	2250.0
233.6	788.5	2350.0
241.8	810.2	2432.0
250.3	855.5	2450.0
258.8	879.0	2550.0
266.0	903.2	2650.0
273.3	928.1	2706.0
280.8	953.7	2750.0
288.5	979.9	2850.0
296.5	1006.9	2950.0
304.7	1034.0	3050.0
313.0	1049.6	3150.0
321.7		3225.8

Tone coding recognition equipment, at the remote location, can become considerably more complex than merely recognizing sequential audio frequencies and a number of sophisticated tone-encoding systems are available having large information capabilities and able to cope with hundreds of thousands of dialers. See Fig. 7-8 for a typical plug-in Reed type tone decoder. This device will operate with an input signal of less than 50 millivolts at a frequency of 368.5kHz, generated by a tone encoder located in the protected premises. Schematic diagrams for a typical tone encoder and decoder

Fig. 7-8. Motorola plug-in Reed type tone decoder (pencil shows comparison).

system are illustrated in Fig. 7-9. Fig. 7-10 shows a forty-eight-position alarm indicator which is used at a remote monitoring point. This device gives a latching visible, as well as audible, indication of an intrusion at any one of forty-eight separate locations. Such equipment is located at the remote point and will usually indicate, in addition to the location of the premises, the time and date of the intrusion as well as the intrusion point (vault, stockroom, main door, etc.). In the even more sophisticated remote tone-encoded installations, this data is actually printed out and forms part of a permanent record. In the simpler installations, an audio-frequency sensing circuit such as a Reed relay is connected across the incoming telephone line. A number of these relays are bridged together up to the total capacity of the system. When the tone frequency associated with a particular relay is received, that relay operates and latches to give a visual and audible indication of the alarm.

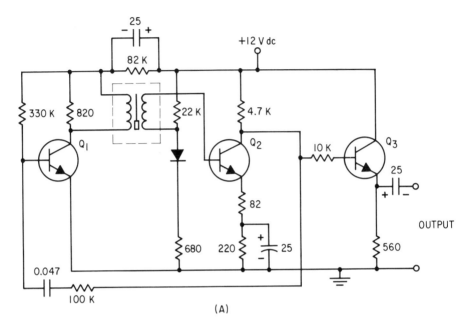

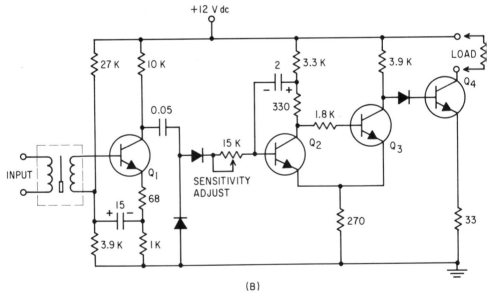

Fig. 7-9. Schematic of tone encoder and decoder units.

Fig. 7-10. Forty-eight-position alarm indicator for remote monitoring.

Telephone Subsets

As previously indicated, both the tape telephone dialer and the digital telephone dialer are usually used in conjunction with the telephone station (subset). Both of these types of dialers can also be used with leased telephone lines, but in this type of service, different dialing techniques are usually employed. When these dialers are used with the telephone station, they can be hard-wired directly to the telephone line; i.e., they can be connected directly to the line conductors without employing interface equipment supplied by the telephone company. In some locations, the telephone company may permit this type of hard wiring. However, in most cases, the telephone company will require the use of an interface device referred to as the KS-20445 L1, L2, or L3 telephone coupler. This coupler (see Fig. 7-11) is installed by the telephone company and is rented to the subscriber for a monthly fee of approximately $3.00.

The telephone company's coupler is a sophisticated device whose sole function is to electrically isolate the dialer from the telephone lines so that the service on the telephone line is not degraded in the event of dialer failure.

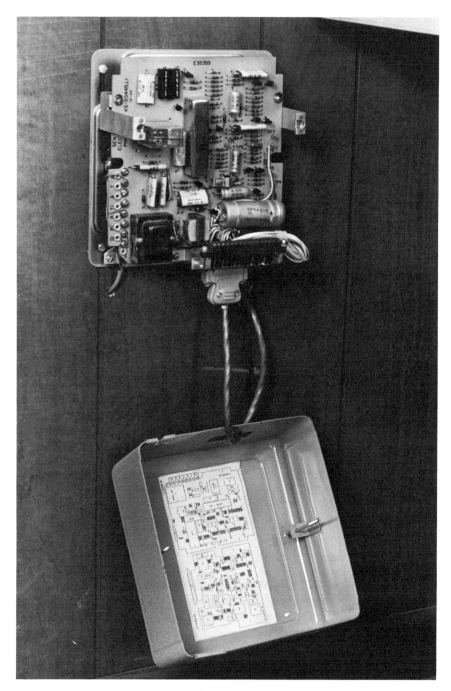

Fig. 7-11. Bell System #KS-20445-L1 telephone line coupler.

Furthermore, the coupler contains circuitry which clips and shapes any signal coming out of the dialer so that it conforms in frequency, amplitude and impedance to the type of signal which the telephone company requires to properly match into its lines. Some telephone couplers require 18 volts dc for their operation at currents of less than 20MA. In addition, the coupler terminates in a special socket to which all connections must be made utilizing the proper mating plug (Cinch Jones DA-19603-403 [plug] and DA-51225-1 [shell]). With respect to the 18 volts which are required by the coupler, many dialers offer this voltage from their internal power supplies. In instances where the dialer used with the coupler does not have an 18 volt supply, a special power supply or batteries must be added.

Although the telephone dialer, when used with the telephone station instrument or subset, is by far the most popular method of generating the remote alarm, its usage has a number of disadvantages. In the first place, if the number to be dialed is busy when the dialer calls in, the message will be lost. This may even be the case with dialers having long repetitious tape programs, since it is conceivable that a line might be tied up in excess of 10 minutes. An even greater disadvantage of this technique lies in the fact that if the intruder cuts the telephone line prior to entering the premises, the remote alarming capability will be completely lost. Both of these disadvantages are overcome through the use of a leased line telephone system which is employed and recommended in professional security applications.

Leased-Line System

In the leased or direct line system, a pair of wires are leased from the telephone company starting at the premises which are to be protected, and terminating at the remote alarm center—be it a central station or a police station, etc. This direct line consists of a pair of leased wires which do not go through the usual telephone switching equipment in the central office and, consequently, provide a metallic circuit for dc operation. A number of remote alarm systems are used with leased lines and all of them have the advantage over the standard telephone line. They will give indications not only when an intrusion takes place, but also when any portion of the remote alarm system is tampered with or cut. The two common types of circuits are the reversing relay system, and the McCulloh loop.

Reversing Relay System

In the current-reversing system depicted in Fig. 7-12, relay RL-1 is located on the protected premises. This relay, which is usually mounted directly in the alarm control equipment or near it, is energized by a suitable voltage source from the control circuit. This relay is wired so that the battery polarity on its contacts will reverse when the relay operates as the result of an intru-

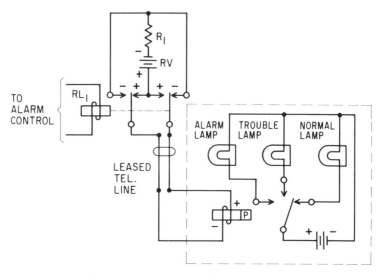

Fig. 7-12. Schematic of ADEMCO #310 remote alarm indicator.

sion. The voltage supplied by battery RV is connected by a leased telephone line to the remote alarm point. Resistor R-1 serves to limit the current flow in case of a short-circuit of the telephone line or in the wiring.

At the remote alarm point, the leased line connects to the winding of the differential or polar relay RL-2. This relay is constructed with a small permanent magnet at the end of its core. When current is not flowing through its winding, the armature of relay RL-2 will cause the center contacts to close, thereby lighting the "trouble lamp." Current flowing through its windings in the direction + to − will cause the relay to release. This will close the rear contacts and the "normal lamp" will light. Current in the − to + direction will cause this relay to operate fully on its front contacts, thereby lighting the "alarm lamp."

The amplitude of the battery voltage, RV, is dependent upon the voltage required to operate the differential relay, RL-2, at the remote alarm point. In addition, the reversing voltage amplitude is also dependent upon the ohmic resistance of the telephone line involved which is a function of the physical length of the line and of the wire size used by the telephone company. In general, battery voltages from 9 to 18 volts will suffice even when the ohmic losses in a telephone line are high. This reversing voltage is usually furnished by batteries or some power supply with standby capabilities.

With reference to Fig. 7-12, under normal condition RL-1 is not energized and RL-2 is energized through a set of contacts on RL-1, but does not operate because of the battery polarity. As long as this condition exists, the "normal lamp" at the remote point is energized. The moment an intrusion is made, relay RL-1 becomes energized and its contacts shift over, reversing the battery polarity on the telephone line. This polarity change causes relay RL-2 to operate, thereby lighting the "alarm lamp." Note that current flows through relay RL-2 both in the normal and in the alarm state, but it operates only when the polarity of the dc overcomes the inherent magnetic force of the permanent magnet in its core.

When the telephone line is cut, the circuit to the coil of the differential relay RL-2 is open and the relay now becomes de-energized with its contact moving to its center on the trouble indicator position. The sensing of polarity reversals and absence of voltage at the remote alarm point need not always be detected by a differential relay. (See Fig. 7-13 for a remote alarm unit.) A number of remote point indicators utilize galvanometers and associated solid-state circuitry to sense these three conditions (normal, alarm and trouble) and give immediate indication of the status of the circuit.

Fig. 7-13. ADEMCO #310 remote alarm unit.

McCulloh Loop System

The reversing-relay circuit, though widely used and very reliable, has the limitation that it requires a separate metallic telephone circuit for each of the premises to be protected. Furthermore, it only offers three pieces of information—normal, alarm, and trouble. Another widely used metallic line circuit is the McCulloh loop circuit (see Fig. 7-14) in which a very large number of protected premises can be wired in series (see Fig. 7-15), all sending the alarm signal to the remote monitoring point via a single pair of leased wires.

132 ELECTRONIC SECURITY SYSTEMS

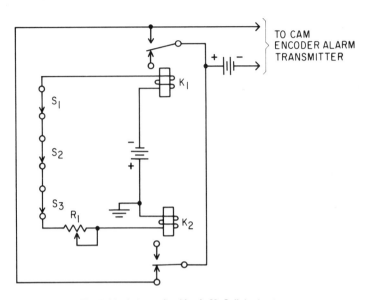

Fig. 7-14. Schematic of basic McCulloh circuit.

In Fig. 7-14, relays K-1 and K-2 are normally energized by the closed loop through sensors S-1, S-2, and S-3. Resistor R-1 is adjusted for minimum current to operate these relays. If any sensor or the line loop is opened, both relays will release and the cam encoder alarm will be activated. If a sensor or the line loop should be grounded, relay K-1 will stay operated, but K-2 will release to activate the alarm circuit.

In the McCulloh system, sometimes called a pen-register system, a multiple cam encoder is located on the protected premises, sometimes actually within the security control equipment. This encoder (see Fig. 7-16) consists of a motor driving a series of cams which, in turn, trigger switches. It is energized from the security control unit. The cams of these encoders or code generators can be adjusted so that they will give a distinct pulse code to identify the premises. Each cam has a notch of different length and a multiplicity of these cams are sequentially staggered on a single shaft. The notches in the cams actuate microswitches which sequentially send a dc pulse down the line. Depending on the number of cams used, an extremely large number of individual identifying codes can be generated. Furthermore, the programming of such a code generator can be set so that the generator, once energized, repeats its message for a preset number of rounds.

As Fig. 7-15 shows, virtually any number of these encoders can be connected in series to a single direct line. This line is terminated at the remote alarm monitoring point and is directly coupled to a decoding device which is

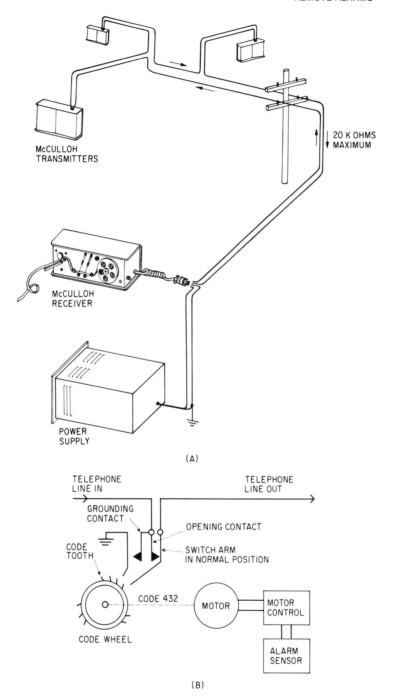

Fig. 7-15. Alarm system utilizing several McCulloh transmitters (A), and (B) Mechanical code transmitter hook-up diagram.

Fig. 7-16. CONRAC #535 Code generator.

called the pen register. The register consists of a large spool of recording paper which is pulled by motor beneath a pen mechanism. The pen mechanism is actuated by the signal developed in the code generator. When an intrusion is made in a protected premises, the code generator immediately sends out its identifying code which in turn is recorded and displayed on the pen register at the remote monitoring point (see Fig. 7-17). Essentially, all that the code gen-

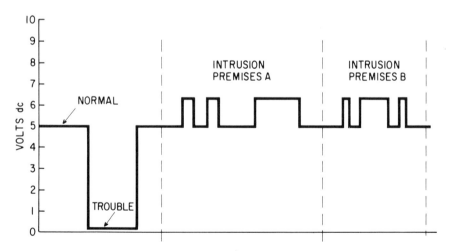

Fig. 7-17. Pen register pulses sent by McCulloh transmitter.

erator does is to raise, during the cam's cycle, the voltage which is normally on the line thereby forming a dc pulse. Each code generator has a unique pulse train and, thus, can be individually identified by the pen register. As in the case of the reversing relay system, if the line is cut, the voltage disappears altogether and the pen register drops to the base-line indicating a trouble condition. Though it is possible with this system to identify the point where an intrusion took place, it is usually not possible to pinpoint the spot where the line is cut because the McCulloh loop system can only indicate if the line has been cut, not the location.

Remote Alarm Locations

Regardless of what type of remote alarm is used (customer or leased line) considerable care must be exercised in selecting the actual remote alarm point. In the early days of security systems, telephone dialers were very popular and were widely programmed into local police stations. Initially, police departments welcomed this innovation in crime prevention and permitted users to program such devices into the police department's main telephone number. As time went by, and as the usage of such devices increased, the main line to police departments would sometimes get jammed as the result of a larger number of automatic dialing devices signalling in. Repeated occurrences of this situation have caused police departments to review their position and many of them now assign a special telephone number solely reserved for dialer service.

In some instances, the abuses of dialers has grown so pronounced that police departments have been instrumental in proposing laws which forbid programming of dialers into the police station. In other cases, police departments have required that each dialer be registered so that they can keep a record of the number of false alarms which, in their opinion, are generated by the device. For example, if such a dialer, in the opinion of the police department, falses more than three times, its card is pulled from the file and calls emanating from this dialer are no longer answered. The understandable resistance, in recent years, of police departments to the acceptance of such dialers has led to a rapid growth of other types of remote answering-monitoring points. Primary amongst these has been the increasing popularity of programming dialers into answering services or central stations which are privately owned. Both of these remote monitoring points can be utilized with customer telephone lines as well as with leased-telephone lines. As a result of these trends, it is advisable to check with the local police department with respect to local dialer requirements.

Because of the great increase in telephone customers and calls, it often is not possible for telephone companies to furnish physical or metallic cable pairs that provide dc paths from the guarded premises to remote alarm points, or to the central alarm office. In many localities, the telephone company can only supply voice-frequency (ac) circuits which preclude the use of dc operations. Consequently, it is becoming increasingly necessary to utilize ac signaling instead of dc for remote alarm purposes.

8
Special Systems

Integrated Systems

In previous chapters, we have seen that the security system consists of an intrusion detector, a control device, and some form of alarm display. We have confined our discussion in these chapters to what is called a "hard wired system"—that is, a system in which all components are interconnected by physical wires. It is possible to eliminate hard wiring by combining all of the generic components in a single enclosure which need only be plugged into an ac outlet in order to obtain the necessary operational power. Such systems are called integrated security systems, and a number of companies have produced devices of this type. These systems are usually supplied in a decorative, tamper-proof steel or wood enclosure and contain some form of volumetric detector, either ultrasonic (see Fig. 8-1), infrared, microwave, or a combination of these, as well as the necessary control circuits, power supplies, standby batteries, and internal siren.

Devices of this type have the tremendous advantage that they can simply be placed in the area which has to be protected, with no further wiring required. All that need be done is to plug the unit into a convenient wall outlet and turn on the key switch, which is usually located on one side of the cabinet. Volumetric protection is then immediately afforded in the area in which such systems are placed. Such systems are ideally suited where fast, temporary security is required, for example, in offices or rooms in which valuable materials may be left for a short period of time.

Fig. 8-1. Mallory integrated security system utilizing ultrasonic volumetric detector and electronic siren.

The advantages of such a system are apparent, in that they offer instant security with virtually no installation time. They also have some disadvantages, and these include the fact that the alarm indicating devices—the siren or the bell—are located within the system itself. If such a system is put into an office in a building, the sound of the siren, no matter how loud it is, is of necessity limited to that room, and the likelihood is very high that the warning signal will not be heard outside the building. Proponents of this type of system feel that a loud siren within a room has such a detrimental effect upon the intruder that he will immediately leave. There is some question, of course, as to whether this position is valid, especially in the light of the fact that the intruder would immediately realize where the sound came from and could take steps to smash the instrument. As previously discussed, devices of this type are usually made of heavy gauge steel, with the siren protected by a strong mesh. Nevertheless, no system, no matter how rigidly constructed, can long withstand an aggressive attack with a sledge hammer or an axe.

Furthermore, devices of this type have limited volumetric detection capability, since they contain a single volumetric detector. Thus, if a series of rooms or a large area, such as a warehouse, is to be protected, a number of such systems are required. Since integrated systems contain all the separate security components, continual duplication of the system, in order to achieve large volumetric coverage, becomes quite expensive.

In order to overcome these disadvantages, it is possible, of course, to attach to an integrated system an external siren mounted on the outside of the building, as well as outboard volumetric intrusion detectors which give greater coverage without duplicating the entire system over and over again. This technique will require hard wiring of the external siren and of the additional outboard volumetric intrusion detectors and the basic advantage of an integrated security system is lost because once again the installer finds himself in the hard wired systems business.

A number of variations exist with respect to the integrated security alarm system which are described below. One type is specifically designed for the urban apartment dweller where the means of entry is usually only through the front door. In such an application, it is not necessary to protect windows or rooms and all one is concerned with is protecting the front door. Integrated alarms of this type are often made in which one-half of the magnetic switch is buried right in the integrated alarm, while the other half is mounted on the front door. These systems usually come in very small boxes and are mounted directly on the front door by means of screws or self-adhesive tape. As soon as an unauthorized person enters through the front door, the alarm trips and a siren tone is emitted.

In order to facilitate ease of installation right on the door frame, integrated alarms of this type must, of necessity, be small and consequently cannot contain large horns or heavy batteries. They are usually powered by a few flashlight cells since ac wiring is also impractical for this type of installation. As a result, alarms of this type emit a siren tone which is not very loud and whose effectiveness is doubtful. Furthermore, the intruder will immediately realize what is happening once he has entered the apartment via the protected door and one smack with a wrench or even stepping on the system after knocking it off the wall will most certainly defeat its purpose. Thus, systems of this type consequently have greater psychological impact on the user than acting as a practical deterrent to intrusion.

Yet another form of integrated alarm systems is the car and truck alarm which is specifically designed to emit a warning sound if an unauthorized person attempts to remove the vehicle. Detectors for this type of alarm could, of course, be magnetic switches in the car door, foil on the windows which do not open, or even a small volumetric intrusion detector in the cab or van. Though these type of detectors are sometimes used, the most common detector for the integrated car alarm is a sensor which is hooked into the vehicle's ignition system. This sensor will detect the spike which appears on the car's ignition system as soon as any electrical device within the car is energized. Such a device might be the dome light in the car which would go on when an unauthorized person opened any one of the doors.

Another method of detection operates on a pendulum principle where a small electric switch is placed somewhere within the car. This switch remains open as long as the car is not in motion. As soon as the car moves or is jarred, the switch trips, causing the alarm system to go into action. Vehicle alarms operate from the vehicle's existing battery, and consequently have complete standby power capabilities.

In summary, integrated security systems do have their place, providing one does not expect them to give the coverage that a hard wired system will provide. Thus, they are ideally suited in small areas, requiring a high degree of security, where the cost of wiring, or the difficulty of wiring, does not permit a more traditional system.

Wireless (Radio) Systems

In previous chapters, we have restricted our discussion to hard wired security systems where the interface between intrusion detectors, control and reporting system consists of physical wiring. The hard wiring technique requires a considerable amount of time and skill on the part of installers, especially in view of the fact that certain applications call for concealed wiring and involve snaking the cables through complex paths. In order to overcome the disadvantages of hard wiring, two wireless (radio) techniques have been developed whereby the signal from the intrusion detector is sent back to the control without the usage of new wires.

The first of these techniques is a wireless system in which a radio-frequency signal is generated at the detection point and sent back, through the air, to a receiver which is located at the control point.

The second wireless technique also generates a high-frequency signal at the detection point, but sends the signals back to the control point by utilizing existing electric wiring.

Wireless Air Systems

With respect to wireless devices, the Federal Communications Commission has allocated certain frequency bands, including 220MHz, for the purpose of transmitting signals of this type. In addition to setting frequency requirements for this type of service, the FCC also limits the transmitted field-strength which such devices generate. (As previously pointed out in Chapter 3, the regulations are quite similar to those applicable to near field volumetric intrusion detectors.) The propagation characteristics at 220MHz are very close to the frequencies employed by the television broadcast channels, and have essentially the same properties; that is, signals of this frequency will easily penetrate, with some attenuation, all substances except metal, such as steel, tin, sheet metal, wire lath, etc.

Furthermore, since the FCC limits the amount of power which these devices can radiate, the effective operational distance over which such pure wireless (radio) systems will function (that is, the distance between the transmitter and the receiver) is limited to a straight unobstructive line-path of about 200 feet. Even this figure represents an optimum condition and assumes virtually no intervening signal reducing objects, such as reinforced concrete walls, cinder block, etc. to say nothing of steel shelving in closets, wire lath, metal partitions, etc. Moreover, the 200-foot figure also assumes the very best, most sensitive, modern, state-of-the-art receivers. Very little can be hoped for in terms of increasing the effective distance, since the sensitivity of most receivers is currently near the theoretical limit. The only increase in distance could be achieved through a change in FCC rules, allowing for higher transmitter output, and this seems extremely unlikely, especially in the light of the fact that continued emphasis is being placed on lessening electromagnetic pollution.

If wireless devices utilized a simple continuous carrier wave (cw) at the assigned frequency of 220MHz, a very high false alarm rate would, no doubt, result. Utilization of a pure carrier wave would mean that any radio frequency signal containing a 220MHz component would trip the receiver and cause a false alarm. This is especially true because this frequency is shared with other licensed services, such as the wireless garage door opener. Also, many powerful radio-frequency signals, not necessarily at 220MHz, have harmonic relationships to the 220MHz signal so that they can easily interfere with the receiver which trips in the presence of a continuous wave (cw) signal. In order to limit the false alarm possibilities of wireless (radio) alarm systems, the 220MHz signal is modulated (coded) in a number of ways, so that the receiver will only trip when a properly coded signal is received. The most common ways of modulating include (see Fig. 8-2):
1. Amplitude modulation (AM)
2. Frequency modulation (FM)
3. Frequency shift keying (FSK)
4. Pulse code modulation (PCM)

Amplitude Modulation

In *amplitude modulation,* the simplest method of coding, the amplitude of the 220MHz carrier is modulated in accordance with a specific single audio frequency, or a multiplicity of audio frequencies. In such a system, the transmitter emits a 220MHz carrier modulated with a 1,000Hz audio tone. The receiver, which is located near the control, is very similar to the ordinary AM radio, except that it is tuned to 220MHz and contains a very sharp audio filter which rejects everything except the 1,000MHz audio tone with which the carrier is modulated. As soon as the 1,000Hz tone is received, the alarm switch in the control is activated. The disadvantage of this simplistic amplitude modulated system lies in the fact that all noise as well as many man-made radio signals are of the amplitude-modulated type. Furthermore, if the

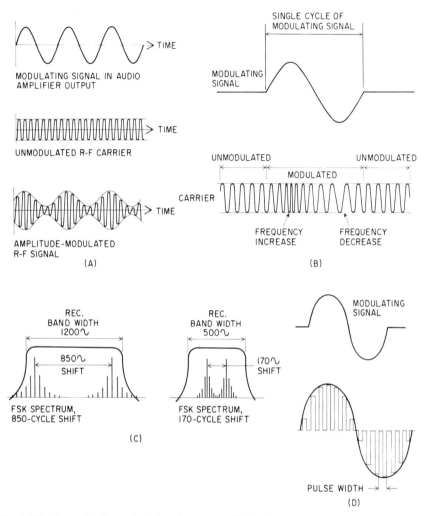

Fig. 8-2. Carrier modulation methods in radio systems. (A) Amplitude modulation (AM). (B) Frequency modulation (FM). (C) Frequency shift keying (FSK). (D) Pulse code modulation (PCM).

noise contains a 1,000Hz audio component—which is quite likely—a false alarm will result. In order to decrease false alarms, manufacturers often employ a series of sequential tones, for example, 1,000Hz followed by 800Hz, followed by 2,800Hz. The statistical possibility of a false alarm with a three-tone sequentially coded amplitude modulated carrier is, of course, considerably less than that of a single tone encoded carrier.

Frequency Modulation

The *frequency modulation* method (FM) is much more widely used than the amplitude modulation method (AM) since it is less susceptible to noise, has more reliability, and a very low false alarm rate. As previously shown, in the amplitude modulation method, the amplitude of the carrier is varied in accordance with the amplitude of the modulating signal. In the frequency modulation method, the frequency of the carrier is varied in direct proportion to the amplitude of the modulating signal. In the absence of any modulating signal, an FM transmitter would, of course, emit the unmodulated carrier (CW) which could be at 220MHz. Let us assume that this FM carrier is now modulated with a 1,000Hz audio tone having a peak-to-peak amplitude of 2 volts. This means that the positive portion of the modulating signal reaches 1 volt, and the negative peak of the modulating signal likewise is 1 volt. A typical frequency modulated system might be designed in which each volt of modulating signal swings the carrier 2,000Hz from the unmodulated frequency of 220MHz. Thus, in the example cited, the carrier would swing up from 220MHz in the amount of 2,000Hz or to 220.002MHz when the positive peak of the modulating frequency was reached. The carrier would swing down 2,000Hz from the basic frequency of 220MHz to 219.998MHz when the negative peak of the modulating signal was attained. The total amount of the signal swing of 4,000Hz is called the FM deviation.

In typical FM security systems, deviation could be as little as 1,000Hz to as much as 30,000Hz, depending upon the amount of information which the FM system must transmit. The receiver in an FM system is designed to accept only signals whose frequencies vary rather than the amplitude in an AM system. Since noise is amplitude modulated, good FM receivers simply do not respond to such adverse signals and consequently will not trigger on them. Furthermore, an extremely low false alarm rate can be insured in an FM system if the modulation signal consists of a series of tones rather than a single tone.

Frequency Shift Keying

Frequency shift keying (FSK) is still another method of generating pure wireless (radio) signal, somewhat similar to FM, and also not prone to false alarming. In frequency shift keying, the carrier is transmitted continuously and an audio tone is used to shift the carrier frequency by some fixed amount. In a typical system, called an audio frequency shift keying (AFSK), the 220MHz carrier is always shifted downward by 850Hz in the presence of the modulating signal. As can be seen, in such a system the amplitude of the modulating signal does not control the amplitude of the carrier (as in AM). Furthermore, if the amplitude of the modulating signal varies, it does not proportionally vary the frequency of the carrier (as in FM). Thus, in the AFSK system, the presence of the modulating signal simply shifts the frequency of the carrier by a predetermined amount. The receiver looks for this shift and triggers the control when it occurs. Once again, noise and most man-made signals do not exhibit FSK characteristics and, therefore, this method of encoding has a very low false alarm rate.

Pulse Code Modulation

Another modulation technique designated *pulse code modulation* (PCM) has been perfected in recent years. It utilizes binary pulses to carry the speech or other signals instead of modulating the carrier as in amplitude (AM) and frequency (FM) modulation as previously described. This modulation method, no doubt, will be used more often in alarm systems in the future.

Battery Power

It must be remembered that regardless of the type of the modulation which is used, the practical transmission limit of wireless (radio) systems under ideal conditions is less than 200 feet, even with the best receivers. Furthermore, since the transmitter is located in the spot where the intrusion detector is, some form of power must be available to the transmitter which does not require wiring. Radio transmitters of the type discussed function in conjunction with traditionally normally open (NO) or normally closed (NC) detectors, including magnetic switches, foil, step mates, etc. They all contain small batteries to furnish the power during the transmitting cycle. The transmitter is only on while the transducer is violated during an intrusion and these transmitters generally do not draw powers in excess of 10 mA.

If, however, they are used with magnetic door switches, they will also be on when the door is opened and closed during normal usage or stay open during normal usage. Thus, it must be remembered that wireless (radio) transmitters need to have their batteries replaced from time to time. It is difficult to predict just how long a battery will last in a wireless system. Most of these transmitters use a standard 9 volt transistor battery and these may last from as little as 1 month to as much as 1 year, depending on the number of times and length that the point is violated. It is impossible to tell how much life there is left in a battery in such a transmitter and great care must be taken to replace them periodically.

Carrier Transmission

In addition to the wireless (radio) systems utilizing energy whose frequency and strength is controlled by the Federal Communications Commission, there is another form of carrier transmission which uses the existing ac power wiring. This type of carrier system has a number of advantages as well as disadvantages over the radio approach. In carrier transmission, the carrier coding technique (modulation) is similar to that in radio systems using AM,

FM, FSK, and PCM modulation. However, instead of the signal being radiated through an antenna, as is the case in radio transmission, the modulated signal is impressed on top of the 60Hz ac line and then is fed through the ac power line from the transmitter back to the receiver.

The advantage of such a system lies in the fact that at the transmitter point no batteries are needed since the system has to be plugged in and consequently takes its power from the ac line. (If standby power is required, in the event of ac power failure, a battery will, of course, be needed.) Note that in a system of this kind, transmission can continue even though ac power is no longer present, since the power lines are merely used as hard wires.

Furthermore, since a system of this type involves interface wiring (in other words, it does not transmit by the traditional methods of radiation) it does not fall under the control of the FCC providing it does not radiate excessively from the house wiring. This means that more powerful transmitters may be used and theoretically longer distances can be covered. It must be remembered that in a carrier system, the energy is transmitted over the electric house wiring which is extremely inefficient at the higher frequencies since it was designed to carry only 60Hz. Consequently, if systems of this type use frequencies as high as 220MHz, the losses through the ac power wiring would be enormous and the signal would travel for only a few feet.

Thus, the operating frequency of power-line carrier transmitters is much lower and may be anywhere from 40kHz to 400kHz (just below the broadcast band). As previously indicated, the range of such a system theoretically can be quite large. As much as 2,000—3,000 feet of standard ac power wiring would easily carry such a signal provided the wiring was not interrupted. This is, of course, a theoretical condition and in actual practice a number of devices are interposed in and between the power line carrying the carrier signals. These devices include power transformers, as well as the continually changing load impedance which appears across power lines as electrical devices are turned on and off.

Each one of these devices tends to severely attenuate the carrier signals which are impressed across the ac line, thereby lowering the range of the carrier transmissions. This is especially true for power transformers which are, of course, designed to pass 60Hz only and severely attenuate carrier signals, especially in the higher frequencies bordering on the 400kHz region. It is usually difficult, if not impossible, to get a carrier signal through a system in which a power transformer is interposed in the line.

Another factor which limits the effective length of carrier transmission is the extensive noise pulses and extraneous signals which are very often found on power lines. These signals result from spurious noises which are generated by electrical equipment normally connected to the line. Such devices include motors, refrigerators, fluorescent lights, typewriters and adding machines, and virtually every electrical appliance. These devices generate pronounced spikes and wave distortions on the ac line and tend to trip the receiver, even though sophisticated modulation techniques, as previously outlines, have been employed at the transmitter end. Thus, in summary, carrier transmission is a viable method for the transmission of security signals without the use of additional hard wiring. However, the effective length over which such systems will work is very difficult to predict and must be tested, on the spot, by the installer. In some buildings, conditions may be so bad that carrier systems may not work for more than 20 or 30 feet and must consequently be abandoned as a technique.

Closed-Circuit Television

Closed-circuit television, although not primarily used for security applications, is of course very useful in security work. A wide variety of companies make standard closed circuit television systems, and these are available for as little as $250.00 for a camera and a monitor to over more than $1,000.00 for systems in which the camera is mounted on motor driven swivels which continually scan the area to be protected while displaying the signal at some remote point with the additional capability of permanently recording the signal on video tape or film.

Thus, a typical closed-circuit TV system would consist of one or a number of television cameras which, by means of a coaxial cable, are fed to a manual or automatic switcher which is located at some remote point. The function of this switcher is to manually or automatically select a specific camera and then display the signal seen by the camera on a monitor. In the automatic switching mode, switchers automatically jump from camera to camera, pausing about 5 to 10 seconds at each camera. More sophisticated cameras may be equipped with automatic scanners which are preprogrammed to move the camera through a predetermined area. Furthermore, a wide variety of lenses are available for such cameras offering zoom capabilities, as well as a variety of viewing angles under various lighting conditions. Generally speaking, closed-circuit television systems in security applications do require that a human operator be placed at the monitor end to continually watch the display.

In recent years, a number of automatic monitoring systems have come into use which do away with the human observers. The simplest of these is a photoelectric system where the photoelectric cells are held to the monitor screen by rubber suction cups. In such a system, up to four photoelectric cells are attached to the monitor screen in such a way that a straight line between each cell circumscribes the area of interest. These photocells actually form a bridge circuit which looks at the light level as it appears in the picture on the monitor screen. If an intruder moves through this scene, the light level at one of the photocells looking at the scene will change, the bridge will unbalance, and an alarm will be triggered. In the event the lights go out or a voltage fluctuation on the power line causes the lights to increase in brightness, such a change simultaneously affects all four photocells and would consequently be disregarded in terms of an alarm situation.

An even more sophisticated system which has recently become available consists of a special signal processor which is electrically interposed into the coaxial cable line between the camera and the monitor. This sophisticated device actually examines and stores the TV scanning signal for each particular scanning frame. The circuits are designed so that if changes take place in the scanning signal from one frame to another an alarm is triggered. This system, just like the photoelectric cells, will disregard changes which affect the entire frame, such as an increase or decrease in the total light level.

The closed-circuit televisions which we have discussed all require some form of lighting. In recent years, a number of sophisticated television systems have become available which can monitor a scene in semi-darkness or even in total darkness. These systems use special camera tubes and elaborate electronic image intensifying systems. Furthermore, they very often employ cameras which are sensitive to the infrared portion of the spectrum and consequently do not need visible light. Closed-circuit television cameras of this type are available and furnish usable pictures where light intensities may be as low as .02-foot candles.

9
Installation Of Security Systems

To a large degree, the success or failure of a wired security system is dependent upon the care and expertise which was applied in installing the system. Skimpy and hurried installation procedures will produce security systems which work for a few days or weeks, and then rapidly commence to false or totally break down resulting in expensive service calls. The importance of conservative, careful installation procedures cannot be overemphasized. Such procedures may be a little bit more expensive in terms of time used and parts consumed, but in the long run will make for a satisfied customer and a minimal call back situation. The basic tools which are considered mandatory in making a sound security installation as listed below:

1. A small portable quality volt-ohm meter (VOM), such as the Simpson 260, 230, or the Triplett 630, 310. (A number of excellent units of Japanese manufacture are also available.)
2. All connections and interconnections should be made with insulated spade lugs which are to be put on with a hand crimping tool. We recommend Smith #2764 spade lug which will take #18 to #22 wire. This insulated lug can be crimped with Smith #4980 crimping tool. A number of other manufacturers make equivalent devices, including Amp, Waldom, etc.
3. An Arrow T-25 staple gun with staples.

4. A set of insulated screw drivers, including small and jeweler's screw drivers.
5. A set of Phillips screw drivers.
6. A set of hollow shaft nut drivers.
7. A pair diagonal 5-inch cutters, Kraeuter #35015 or equal.
8. A pair 6-inch long-nose pliers, Kraeuter #20140 or equal.
9. A 6½-inch combination pliers, Kraeuter #27507 or equal.
10. A pair wire stripper and cutter, Miller Model #100 or equal.
11. A 47-watt small soldering iron, Ungar Model #777 or equal.
12. A 1-lb. spool of Kester Resin #5 solder 40-60.
13. A 25-foot spool of #22 stranded insulated hookup wire.
14. A ¼-inch electric drill.
15. A set of drill bits including special long drills.
16. A roll of Scotch electrical tape, type 33, ¾-inch wide, in dispenser.
17. A box 1-Amp Slo-Blo fuses.
18. A box 2-Amp Slo-Blo fuses.
19. A box 3-Amp Slo-Blo fuses.
20. A box 5-Amp Slo-Blo fuses.

In addition to these tools, certain important universal installation suggestions should be borne in mind.

Cables

Virtually all security systems use a variety of cables from single to multiple conductor. As indicated in Chapter 5 on power supplies, the installer must make certain that the current-carrying capacity of the cable used is enough to handle the devices which the cable feeds. Perimeter cables, as previously shown, carry very little current and, consequently, thin insulated wire will usually suffice (22–28 gauge). However, cables carrying voltages to bells, lights, control stations, or ac outlets, may carry considerable amount of current and heavier wire must be used, especially where long runs are involved.

Figure 9-1 and Table 9-1 show various cable sizes, their resistance, and current carrying capacities over long runs. Once the proper wire size has been established, all cables should be secured to the walls with proper electrical staples and a staple gun. Generally speaking, the practice adhered to by the installers from telephone companies show excellent installation procedures which should be followed by the security installer. Under no circumstances should cables be left unsecured or secured by nails or staples which are not specifically designed for cables. It is also recommended that all interconnections, including cable splices and especially connections to the control,

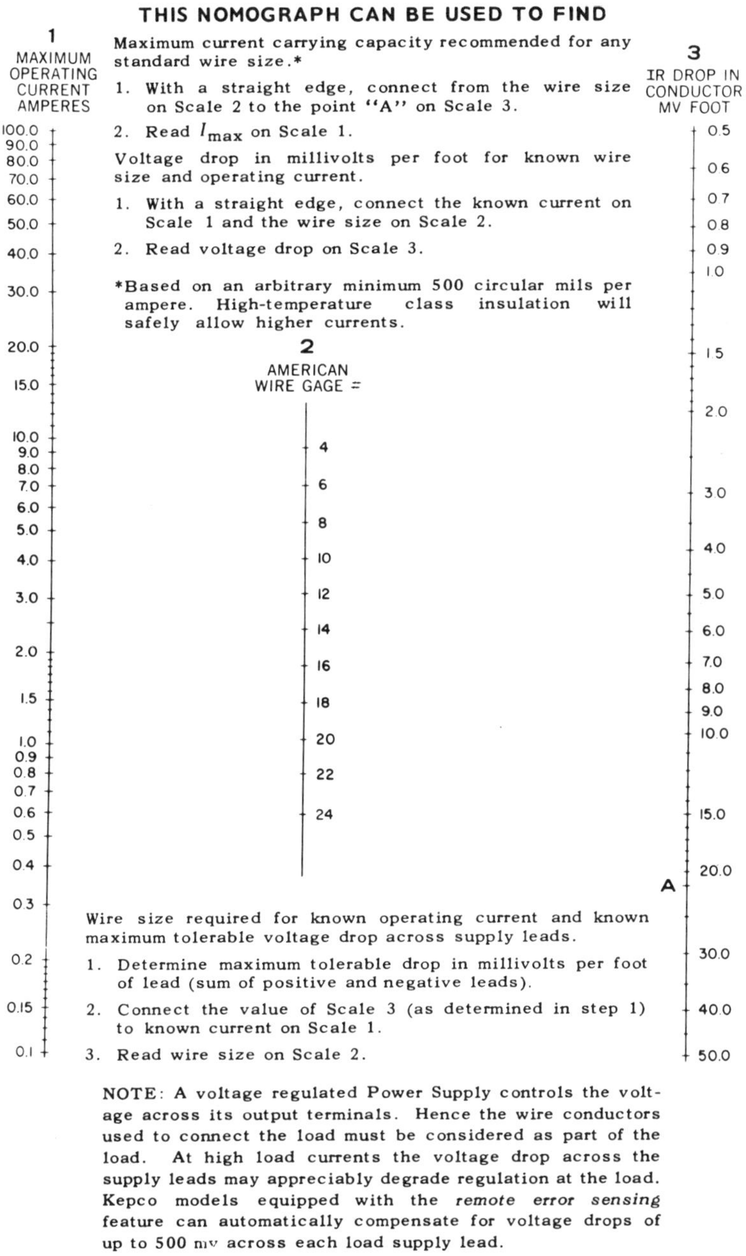

INSTALLATION OF SECURITY SYSTEMS 151

THIS NOMOGRAPH CAN BE USED TO FIND

1 MAXIMUM OPERATING CURRENT AMPERES

Maximum current carrying capacity recommended for any standard wire size.*

1. With a straight edge, connect from the wire size on Scale 2 to the point "A" on Scale 3.
2. Read I_{max} on Scale 1.

Voltage drop in millivolts per foot for known wire size and operating current.

1. With a straight edge, connect the known current on Scale 1 and the wire size on Scale 2.
2. Read voltage drop on Scale 3.

*Based on an arbitrary minimum 500 circular mils per ampere. High-temperature class insulation will safely allow higher currents.

2 AMERICAN WIRE GAGE =

3 IR DROP IN CONDUCTOR MV FOOT

Wire size required for known operating current and known maximum tolerable voltage drop across supply leads.

1. Determine maximum tolerable drop in millivolts per foot of lead (sum of positive and negative leads).
2. Connect the value of Scale 3 (as determined in step 1) to known current on Scale 1.
3. Read wire size on Scale 2.

NOTE: A voltage regulated Power Supply controls the voltage across its output terminals. Hence the wire conductors used to connect the load must be considered as part of the load. At high load currents the voltage drop across the supply leads may appreciably degrade regulation at the load. Kepco models equipped with the *remote error sensing* feature can automatically compensate for voltage drops of up to 500 mv across each load supply lead.

Fig. 9-1. Nomograph of voltage drop across load supply leads (as a function of wire size and load current).

Table 9 – 1. Copper Wire

Wire Size A.W.G. (B&S)	Diam. in Mils*	Circular Mil Area	Turns per Linear Inch† Enamel	Turns per Linear Inch† S.C.E.	Turns per Linear Inch† D.C.C.	Cont.-duty Current‡ Single Wire in Open Air	Cont.-duty Current‡ Wires or Cables in Conduits or Bundles	Feet per Pound, Bare	Ohms per 1,000 feet 25° C	Current Carrying Capacity§ at 700 CM per Amp.	Diam. in mm.	Nearest British S.W.G. No.
1	289.3	83690						3.947	.1264	119.6	7.348	1
2	257.6	66370						4.977	.1593	94.8	6.544	3
3	229.4	52640						6.276	.2009	75.2	5.827	4
4	204.3	41740						7.914	.2533	59.6	5.189	5
5	181.9	33100						9.980	.3195	47.3	4.621	7
6	162.0	26250						12.58	.4028	37.5	4.115	8
7	144.3	20820						15.87	.5080	29.7	3.665	9
8	128.5	16510	7.6			73	46	20.01	.6405	23.6	3.264	10
9	114.4	13090	8.6					25.23	.8077	18.7	2.906	11
10	101.9	10380	9.6	9.1	7.1	55	33	31.82	1.018	14.8	2.588	12
11	90.7	8234	10.7		7.8			40.12	1.284	11.8	2.305	13
12	80.8	6530	12.0	11.3	8.9	41	23	50.59	1.619	9.33	2.053	14
13	72.0	5178	13.5		9.8			63.80	2.042	7.40	1.828	15
14	64.1	4107	15.0	14.0	10.9	32	17	80.44	2.575	5.87	1.628	16
15	57.1	3257	16.8		12.8			101.4	3.247	4.65	1.450	17
16	50.8	2583	18.9	17.3	13.8	22	13	127.9	4.094	3.69	1.291	18
17	45.3	2048	21.2		14.7			161.3	5.163	2.93	1.150	18
18	40.3	1624	23.6	21.2	16.4	16	10	203.4	6.510	2.32	1.024	19
19	35.9	1288	26.4		18.1			256.5	8.210	1.84	.912	20
20	32.0	1022	29.4	25.8	19.8	11	7.5	323.4	10.35	1.46	.812	21
21	28.5	810	33.1		21.8			407.8	13.05	1.16	.723	22
22	25.3	642	37.0	31.3	23.8		5	514.2	16.46	.918	.644	23
23	22.6	510	41.3		26.0			648.4	20.76	.728	.573	24
24	20.1	404	46.3	37.6	30.0			817.7	26.17	.577	.511	25
25	17.9	320	51.7		37.6			1031	33.00	.458	.455	26
26	15.9	254	58.0	46.1	35.6			1300	41.62	.363	.405	27
27	14.2	202	64.9		38.6			1639	52.48	.288	.361	29
28	12.6	160	72.7	54.6	41.8			2067	66.17	.228	.321	30
29	11.3	127	81.6		45.0			2607	83.44	.181	.286	31
30	10.0	101	90.5	64.1	48.5			3287	105.2	.144	.255	33
31	8.9	80	101		51.8			4145	132.7	.114	.227	34
32	8.0	63	113	74.1	55.5			5227	167.3	.090	.202	36
33	7.1	50	127		59.2			6591	211.0	.072	.180	37
34	6.3	40	143	86.2	62.6			8310	266.0	.057	.160	38
35	5.6	32	158		66.3			10480	335	.045	.143	38-39
36	5.0	25	175	103.1	70.0			13210	423	.036	.127	39-40
37	4.5	20	198		73.5			16660	533	.028	.113	41
38	4.0	16	224	116.3	77.0			21010	673	.022	.101	42
39	3.5	12	248		80.3			26500	848	.018	.090	43
40	3.1	10	282	131.6	83.6			33410	1070	.014	.080	44

* A mil is .001 inch.
† Figures given are approximate only: insulation thickness varies with manufacturer.
‡ Max. wire temp. of 212° F and max. ambient temp. of 135° F.
§ 700 circular mils per ampere is a satisfactory design figure for small transformers, but values from 500 to 1000 c.m. are commonly used.

should be made with crimping spade lugs utilizing any one of the commercially available crimping tools. Bare-wire connections under terminal screws will invariably lead to trouble and make for a very sloppy installation. Furthermore, in security installations it is very common to encounter a number of multi-conductor cables. It is important to place tags on each cable indicating the purpose and termination of the conductors. This technique becomes especially important at the control end where a number of these cables may terminate. Finally, testing and subsequent servicing becomes virtually impossible if one is confronted with a rat's nest of wires and cables totally without identification.

Perimeter Devices

Foil

One of the most popular perimeter intrusion detectors utilizes a metallic foil in tape form. This foil is glued around the edges of windows and other members which will break during illegal entry thereby setting off the alarm. Foil of this type is available in a number of widths and lengths. The perimeter loop wires which must be connected to the foil can either be soldered to the foil or can be attached to the foil with special connectors supplied by the various manufacturers (Figs. 9-2 and 9-3). Proper foiling is a technique which will become very easy only after a certain amount of practice. It is, consequently, strongly urged that prior to doing the first foil job, a practice run be attempted utilizing scrap glass. In placing foil, remember that the windows involved must be very tight in their frames. Any loose rattling window will tend to break the foil as it runs over the stripping. Furthermore, the windows must be dry, since the foil will not adhere to wet or sweating surfaces. It is also recommended that any surface which is to be foiled be first cleaned with a good detergent and dried.

The foil path should be laid out with a grease pencil or tailor chalk on the reverse side of the glass. The foil run should, of course, occur in that portion of the window which must break if an illegal entry is made. A number of patterns are possible, but the most common pattern is a rectangular circumferential pattern with the foil running approximately 6 inches from the outside edge of the window. Some foils are currently available which are self-adhesive, but a majority of foils are still applied with a combination glue and foil protector, which usually is some form of varnish. Actually, any good grade of varnish can be used for this purpose. However, it is recommended that the special foiling fluids which manufacturers recommend be used instead of ordinary varnish. After the guidelines have been drawn, a coat of the foil varnish is applied along the guidelines, on the opposite side of the window, with a brush about ⅜ inch wide. After the varnish has been applied, the installer should wait until it becomes tacky before actually applying the foil.

154 ELECTRONIC SECURITY SYSTEMS

Fig. 9-2. Installation of door cord to window foil for perimeter security protection.

INSTALLATION OF SECURITY SYSTEMS 155

Fig. 9-3. Typical installations of magnetic switches and window foil.

Foils are usually supplied in the form of rolls and it is much easier to build a small foil dispensing stand made up of a nail and a couple pieces of wood so that the foil can be rolled out easily without spilling or unraveling. It is best to apply the foil in one continuous strip, starting at the edge of the glass. It is a good idea to leave a few inches of foil at the edge as an additional tab. This tab is eventually folded back on itself and gives that point a double foil thickness for circuit takeoff purposes. A dab of varnish can be put underneath the tab to hold it secure to the main foil section. In applying the foil, a squeegee technique should be used and the foil should be smoothed onto the glass using some cardboard or a small piece of round solid plastic tubing as a squeegee. The foil should be applied smoothly and any lumps or air bubbles should be removed with the squeegee.

It sometimes becomes necessary on certain windows to pass the foil over metal frames and braces. These metal parts are usually electrically grounded and the foil must be insulated from them. To achieve an insulated crossover, a piece of standard electrical tape should be placed beneath the foil where it crosses. The tape is self-adhesive, but it is recommended that varnish be used in addition to hold the tape to the crossover point. Just enough insulating tape should be used to prevent the foil from touching the metal. If too much insulating tape is used, the buildup in thickness on the window may get excessive and the whole crossover will be ripped apart during window cleaning. The actual crossover can be done with tinned brass or copper shim stock. A piece of this material is cut and shaped in the configuration of the crossover, leaving about ¼ inch on either side of the crossover extending over the foil on the window. It is generally good practice to make sure that this crossover shim is thoroughly tinned with solder, especially the section which will come in contact with the window foil. Varnish is then applied to the insulating tape on the crossover and when it becomes tacky the tinned brass or copper is pressed on top of it. Utilizing a soldering iron and working very gently, never applying too much heat for too long a period, the shim crossover is then tacked to the foil on either side of the window. Once this has been done, protective varnish should be applied over the entire crossover and when that has dried, electrical tape should be used to protect the complete crossover.

At the termination points of the foil, in other words the beginning and the end, the wires from the perimeter circuit must be hooked in. This can be done in a number of ways, the simplest of which is to make a foil connector using tinned shim stock. The shim stock is cut and shaped to the form of the termination point. A wood screw is used to fasten one end of the shim stock into the window frame or molding while the other end of the shim stock is gently soldered to the foil. Perimeter loop connections can then be made with the wires soldered to the shim stock at the screw end. The entire connection section can then be covered with protective varnish followed by a layer of electrical tape. Although this final hookup technique works, it is recommended that the foil be terminated by means of any one of the recommended foil connectors which are manufactured by a number of companies. Some of these are of the pressure type which do not require soldering. Once the foiling job has been completed, it is recommended that the entire foil be cleaned with benzine, removing dirt and excess varnish. After this has been done, another coat of varnish should be applied over the entire foil system. In repairing foil breaks, small pieces of foil can be used as a patch. In applying the patch, care must be taken that good electrical contact is made between the broken sections.

Magnetic Switches

Another widely used perimeter intrusion detector is the magnetic switch. As discussed in Chapter 2, a number of versions of this device are available employing Reed switches, as well as sliding-contact switches. Various magnetic switches, from different manufacturers, will function over different switch gaps. In purchasing these switches, the effective operating gap can be determined by moving the magnet section in front of the switch section and listening for the switch click. Some of the better magnetic switches will operate even when the magnet is more than 1 inch away from the switching member, while the lower priced units will fail to function if the gap between the magnet and the switch exceeds ½ inch. Magnetic switches can be used in a multitude of applications and even though they may operate with wide gaps, it is advisable to keep the gap as small as possible when the switch is in the normal (unviolated) position.

When these switches are used on doors, they are generally mounted on the top or on the side that opens. This type of mounting makes the switch least conspicuous and offers the highest degree of sensitivity to an intrusion. The magnet part of the switch, since it has no wires connected to it, is always mounted on the door while the contact section is mounted on the frame (see Fig. 9-4). Magnetic switches should never be installed on the hinged side of the door, since with that type of installation the gap between the magnet and the contact section may not become wide enough to permit system triggering. Furthermore, with that type of installation an intruder could open the door, reach in, and by means of an auxiliary magnet or jumper wires defeat the entire system.

Magnetic switches can, of course, also be used for sliding windows as well as for casement and awning windows. When used with windows, two magnets can be employed so that a window can be left partially open and still maintain the integrity of the perimeter loop. In this type of installation, the window is open to the amount required for ventilation and a second magnet is then mounted adjacent to the contact section of the switch. If this type of installation is done on double-hung windows, it is preferable to use the lower window for ventilation since the usage of a second magnet on the upper win-

158 ELECTRONIC SECURITY SYSTEMS

Fig. 9-4. Magnetic switches equipped with manual shunt lock to permit exit via a protected door.

dow would limit the amount that the lower window can be opened. Obviously, when installing magnet switches in an application where ventilation is required, the window opening should not be made so large that an intruder can climb through without triggering the system.

Step Mats

Step mats, which trigger when weight is placed on top of them, are made in a variety of shapes and sizes including continuous long runners, small rectangular mats, and very narrow strips, and circular mats which are not much larger than a quarter. These devices are electrically opened and closed in the violated state. The mats are available in very thin sections. Their widest use is underneath carpeting. These mats should always be placed directly on the floor with the rug and padding on top—never between rug and padding. When used in applications where moisture is involved, such as on concrete floors with rubber runners, it is a good idea to place a sheet of plastic or rubber between the runner and the concrete floor to prevent moisture from entering the step mat switch circuitry.

Step mats are also available in strip form having a width of approximately ½ inch. This type of step mat is ideally suited as a windowsill trap since it is quite flat, inconspicuous, and will electrically close even under the slightest pressure. When used in this fashion, the windowsill step-mat may have a pressure sensitive backing or it is glued directly to the sill.

In terms of actual construction, most step mats consist of flexible sheaths, such as rubber, covered on the inside surface with a conductive material, such as copper. In normal use, the two sheaths are separated and electrical contact is formed when the sheaths touch each other as a result of someone stepping on them. Thus, step mats are only available in the normally open configuration and must consequently be used in systems which accept normally open (that is, closure upon violation) devices.

Furthermore, the number of closures which a step mat can take over a period of time is limited. It is difficult to make life predictions, but step mats must from time to time be replaced, especially in areas where there is considerable traffic. The best way to test a step mat is through the use of an ohmmeter which should read virtually zero resistance when someone is stepping on the mat, and near infinite resistance when the mat is clear.

Vibration Contacts

Another form of perimeter intrusion detector is the vibration contact. This detector usually consists of a phosphor bronze leaf with a metal counterweight and contact point at one end. This leaf is mated with a fixed contact. The distance which the leaf is from the fixed contact is usually adjusted by means of a setting screw. When this type of intrusion detector is used, the leaf vibrates as a result of the vibration in a structural member which is due to forced entry generated by saws, chisels, hammers, etc. As a result of this vibration, the detector makes momentary contact which is enough to trigger the alarm system. Usually these types of contactors may be used on iron, brick, cement block, or plywood. Their vibrating leaf with its weight is designed in such a manner that it will not vibrate as a result of normal building motion. The adjustment of the vibration contact is usually made in terms of grams pressure as set by the leaf setting screw.

Most vibration contacts have gram readouts on their setting screw and these will vary with different types of usages. Some manufacturers give specific gram settings for their vibration contacts when used on various structural members. Generally speaking, settings between 5- and 20-gram pressure will cover every conceivable situation. In mounting vibration contacts, it is recommended that they be placed where the vibration will be most severe during forced entry. An excellent place to mount these contacts is on the firring strip on walls and ceilings. These strips tend to act as sounding boards and will cause the vibration contact to close. In installing these contacts, they should be individually tested with a continuity tester or ohmmeter and their leaf pressure adjusted to an optimum point. To test the setting, light hammer blows 5 to 10 feet away from the detector may be used as a signal source.

Perimeter Interconnecting Devices

Whenever cable connections are to be made between a stationary point and a moving point, a special flexible cable with fittings called a "door-cord" is used. Primary uses of the door-cord are between door mounted components (such as shunt keys, controls, indicators) and the fixed wiring in the security system. The door-cord can also be used to connect window foil on movable windows to the perimeter loop.

In utilizing door-cords, they must always be installed on the hinged side of the door. The door-cord's actual position on the hinged side depends on the wiring of the system. In installing door-cords, always mark off the locations of the mounting holes for the cord connecting blocks with the door in such a position that it will require the maximum extension of the door-cord. Consequently, if the cord is slack when the door is open and fully extended when the door is closed, the mounting holes for the block should be marked off when the door is closed. If this procedure is not followed, the door-cord or the associated woodwork can be easily damaged.

When using a door-cord in conjunction with window foil, the window foil should, of course, be terminated on the hinge side of the window. Generally speaking, such usage of a door-cord is intended for casement windows or French doors and windows. With such windows, the standard door-cord having a length of 12 inches will suffice. Windows which hinge in the center and double hung windows can also be used with 12 inch door-cords. The length of such cords, however, would limit the amount of opening which is possible in a window and a protective window stop should be mounted to the window frame. Where a full opening of such windows is required, retractable, coiled door-cords are available which stretch out to 36 inches and retract down to 12 inches permitting full opening of the window.

Volumetric Intrusion Detectors

In installing volumetric intrusion detectors, it is of prime importance that the instructions supplied by the manufacturer be followed to the letter. As explained in Chapter 3, these devices are quite sophisticated and will most certainly cause false alarming if they are not installed in accordance with the installation instructions. In using such devices, some very basic universal installation recommendations can be made. When installing all volumetric intrusion detectors try to:

1. Mount with maximum rigidity.
2. Keep sensitivity controls as low as possible.
3. Where large areas are to be covered, use more than one detector rather than raising sensitivity.
4. Where external power supplies are needed, use only highly filtered, well-regulated electronic supplies.
5. If device contains relay, do not switch large loads (more than 5 watts) through relay contacts.
6. Follow manufacturer's instruction literature carefully. Do not make adjustments other than those permitted.

Some practices to avoid during installations are outlined in Table 9-2.

Table 9-2. Volumetric Intrusion Detector Installation Precautions

VOLUMETRIC INTRUSION DETECTOR	TRY TO AVOID
915MHz and 2.5 and 10.5GHz	Environments with strong RF fields; positioning on exterior walls where trees and traffic might trigger; installations where detectors are close (less than 30') to each other; Mounting near (2' or less) flourescent lights; mounting near (2' or less) large metal objects like filing cabinets and desks.
Ultrasonic	Installations in environments utilizing hot air heating; mounting devices near air conditioning or ventilation outlets; mounting devices near (6' or less) equipment emitting ultrasonic energy (telephone bells, radiator valves).
Sonic	Installations with high ambient noise levels.
Infrared	Installation of devices looking into heat sources (rising sun, radiators, incandescent lights).
Stress	Structural members which flex with temperature changes; buildings which shake in wind gusts; structural members which vibrate with thunder.

As previously pointed out, many of these detectors work on the Doppler principle. This actually means that such devices are very sensitive to motion. Such motion can be generated by the intruder or by the intrusion detector itself in the event that it is not securely mounted. The importance of securely mounting volumetric intrusion detectors cannot be overemphasized. (Refer to Fig. 9-5 which depicts a photoelectric cell system mounted on a rigid bracket with a guardrail to protect the system. Dry cell standby batteries are mounted in a tray below the photoelectric system.) If they are loosely mounted and rattle, ever so slightly, as a result of vibration the net effect will be a Doppler shift which in turn will cause an alarm. It is also good practice to place volumetric intrusion detectors as far away as possible from any structure or device which might generate a false alarm signal.

For example, all volumetric intrusion detectors which operate in the ultrasonic mode, have a tendency to trigger on spurious ultrasonic signals which may be generated within the environment. As a result, such intrusion detectors may trigger on the harmonics of telephone bells, steam and air hiss-

162 ELECTRONIC SECURITY SYSTEMS

Fig. 9-5. Installation of photoelectric cell with guardrail.

ing, noises from jet airplanes, and other spurious signals which are rich in ultrasonic frequencies. Consequently, it is good practice to rigidly mount such detectors as far away as possible from such spurious sources. Furthermore, if such a source, for example a nearby telephone bell, continues to be troublesome, steps can be taken to lower the harmonic content of the false alarm causing device. In the case of the telephone bell, this can be achieved by deadening the bell through the use of a piece of adhesive tape underneath the telephone clapper. To reduce radiator hiss, the radiator valve can be readjusted by a piece of cloth wrapped around the valve escape port. Furthermore, ultrasonic volumetric detectors may also trigger on moving masses which may not be in the form of an intruder, but still generate false alarm causing Doppler signals. Such moving masses can be swaying venetian blinds, air conditioning, and heating ducts which flex back and forth (even if slightly), blowing curtains, moving machinery (such as fans and fly wheels), and even extreme hot and cold air currents. In mounting ultrasonic volumetric intrusion detectors, they should be kept away as far as possible, and never directly faced into, this type of condition.

Volumetric detectors which work in the radio frequency portion of the spectrum (915MHz, 2.5GHz and 10.5GHz) must, of course, also be rigidly mounted since they also generally operate on the Doppler principle. These devices work on electromagnetic energy and are not affected by sound or ultrasonic signals, such as hissing and telephone bells. However, the energy which they emit is reflected just like ultrasonic energy and, consequently, they tend to trigger falsely if they see movement emanating from venetian blinds, air conditioning ducts, etc. As a matter of fact, the energy emitted by these devices will pass through structural members such as brick and cinder block, especially in the case of volumetric detectors operating at 915MHz (this condition is not true for ultrasonic signals). Consequently, these devices may falsely trigger if they "see," on the outside of the protected premises, large moving reflected masses, such as trailer trucks, trains and cars, even swaying trees.

Infrared volumetric intrusion detectors, likewise should be rigidly mounted since they look for heat as a signal to trigger. They should not be pointed into windows where the sun rises, look into radiators or nearby incandescent light bulbs which might energize ehm when the system is in operation.

With respect to sonic volumetric detection, they should be kept away from devices which may produce loud sounds containing frequencies in the audio spectrum which may trip the sonic device. These might include machinery, and television and radio sets which are turned on, etc. Furthermore, these detectors also work on the Doppler principle utilizing sonic energy as the emitted signal and will consequently trip if they see a false alarm causing signal, such as a moving curtain, a waving chandelier, or a fan which rotates in the breeze.

It is evident that volumetric intrusion detectors, no matter how well they are designed, will have a somewhat higher false alarm rate than perimeter counterparts. On the other hand, as previous chapters have pointed out, perimeter intrusion detectors only protect the point of entry and are easily bypassed by jumping the connection (in the case of normally closed devices) or cutting the wires (in the case of normally open devices). Volumetric intrusion detectors protect three dimensional space and—if properly installed—will trip when someone enters the protected area. Because of the sophisticated circuitry used, they may be somewhat more sensitive to possible false alarms but also have a much higher degree of probability of protecting the premises against violation.

One further installation hint should be noted with respect to virtually all volumetric intrusion detectors. As Doppler systems, they all emit some form of energy at a specific frequency. If a number of these devices are used in a single environment, there is a possibility that one device will "talk" to another, thereby causing a false alarm. In order to limit this possibility, manufacturers of volumetric intrusion devices which can be used in multiples, usually give their device some form of frequency code, such as letters A, B, C, and D. When these devices are used in multiple form, the same frequencies should not be adjacent to each other to preclude overlapping of their coverage patterns. Thus, one would put an A near a D, followed by a B, followed by E, etc. It is most important to follow this recommendation since Doppler devices operating at the same frequency will sweep through each other and have a high likelihood of tripping each other without an intruder actually being present.

Decals

Most security system manufacturers supply with their equipment or as an accessory a warning decal which is applied at the protected premises to some point which is easily visible to the potential intruder. Typical application points include inside windows, near entrance points, on the inside of glass doors, basement windows, etc. These decals are of the self-adhesive type and inform the potential intruder that the premises are electronically protected against intrusion. It is recommended that these decals be liberally applied. If the equipment employed does not include such a decal, they may be purchased separately from various sources.

Special Tools

A good basic tool kit complement was presented in the beginning of this chapter. There are, of course, a number of special tools available which make installation of security systems more rapid and efficient. One of these is a wireless portable soldering iron made by the Wahl Corporation. This small portable iron contains rechargeable batteries and a recharging stand into which the iron is placed when it is not being used. Such a device is almost indispensable in making interconnections, repairs, and hookups, especially in places where it is virtually impossible to obtain electric power conveniently. The iron is only able to solder light joints, but does contain a built-in light so that one can see exactly what one is doing.

Another very useful tool in the installation of security systems is a single or double headphone. Although a headphone certainly does not have the flexibility and accuracy of the VOM meter described in the tool kit list, it can quickly establish the presence or absence of low dc or ac voltages. Headphones are available from virtually any radio parts store and are priced anywhere from $5.00 to $15.00. In using headphones as a continuity tester, the prods are placed across the circuit being tested. If a click is heard in the headphone, the presence of direct current is assured. A hum in the headphone means the presence of alternating current in the circuit. It must be pointed out that such headphones can only be used in low voltage circuits which do not exceed 30 volts.

Security installations use a wide variety of cables from a single- or two-conductor perimeter wire to multi-conductor control cables. As previously indicated, these cables must be neatly stapled utilizing a proper staple gun and staples. In a majority of installations, external wiring will suffice. However, in some instances it may become necessary to snake the cable through walls and floors. For this type of work a special drill called the Diverse Bit is used. This tool actually is a long special drill on a flexible steel shaft. These drills are supplied with a special placement tool and are used to drill holes in blind spots such as rafters and crawl spaces. Any medium or slow speed electric drill can be used with this bit. Furthermore, each bit has a hole in its tip to that a snaking wire can be directly hooked in for pulling out a cable. Bits are available in the auger combination, or masonry type. Furthermore, a special recovery grip can be attached to the end of the bit which includes a mesh grip for grabbing cables in blind spots.

INDEX

INDEX

Alarms (*see* Local alarms, Remote alarms)
Ambient light intrusion detector (*see* Space intrusion detector, photoelectric type)
Ampere-hours, 95
Amplitude modulation, 141-142
Antennas, 45
Audio detection principle, 52
Audio filters, 55
Automatic recycling, 77
Automatic scanners, 146

Batteries, 70-71, 83-84, 89, 93-97
   carbon zinc type, 94
   dry cell type, 93
   Gel-Cell, 94, 97
   nickel cadmium, 94, 97
   rechargeable, 97
   shelf life of, 97
   wet cell, 94
   in wireless systems, 144
Beacon alarms, 112
Bells, 18, 20, 99
Bullet switches, 33

Cables, 150-153, 165
Car or truck alarm, 139
Circuit breaker, 92

Carrier transmission of alarm signals, 144
Central station, 25, 129
Chokes (inductors), 88
Clock oscillator circuit, 121
Closed circuit television, 146-147
   automatic scanners for, 146
   automatic monitoring systems for, 146
   signal processor for, 147
Common mode rejection capability, 54
Compressed air sirens, 110
Continuous power, 84
Control terminal strips, 78
Controls, 15, 65-81
   automatic recycling of, 77
   key switches for, 71-72
   location of, 69
   operation of, 69-71
   panic button, 81
   remote stations, 72-74
   shunt lock in, 72
   supervised loop wiring in, 80
   terminal strips in, 78
   zone selection, 78
Coupler for telephone dialer, 127
Current limiting circuitry, 92

Decals, 164
Decibels (dB), 20

Dialer (Automatic), 22-24, 113-129
  clock oscillator circuit in, 121-122
  coupler for, 127-129
  digital type, 113-129
  line seizure capability of, 116
  tape programming for, 114
  tape telephone type, 114-118, 123, 125
  tone encoded messages in, 123
Differential amplifier, 53
Digital dialers (see Dialers)
Digital processing, 60
Digital time sampling processor, 60-61
Direct line system, 23, 129-135
  McCulloh loop system in, 131-135
  pen register system, 131-135
  reversing relay system, 129-130
Door cord, 160
Doppler principle, 9, 35, 47, 161
Double alarm circuit, 77

Electronic sirens, 21, 99, 106
Electronic time delay mechanism, 73
Exit delay problem, 75

False alarming, 13-15, 47, 48, 53, 57 160, 164
F.C.C. regulations, 39, 41-45, 129
Field disturbance sensor, 41
Filters (audio), 55
Flip flop circuit (shift register), 60
Float charging (of batteries), 93
Foil (metal), 6, 33
  installation of, 153-157
Frequency modulation in wireless alarm systems, 143
Frequency shift keying in wireless alarm systems, 143
Fuse, 91

Geophones (seismic transducer), 51
Grounding techniques, 87
Gunn diode, 40

Hard wired security circuit, 65

Inductors (chokes), 88
Infrared beams, 56
Infrared detector, 12, 163
Installation procedures, 149-165
  cables for, 150-153
  for Doppler devices, 161
  for foil, 153-157
  for interconnecting devices, 160
  for magnetic switches, 157-158
  for step mats, 158-159
  tools for, 149-150, 164-165
  for vibration contacts, 159-160
  for volumetric intrusion detectors, 160-164
Integrated systems, 137
Interconnecting devices, 160
Intermittent power, 84
Invisible beam (see Infrared)

Key switches, 19, 71-72

Latching relay, 17, 74
Leased line system (see Direct line system)
Line regulation, 88-89
Line seizure capability of telephone dialer, 116
Load regulation, 88-89
Local alarm, 19, 21-22, 99-112
  beacons, 112
  bells, 104
  compressed air siren, 109
  electronic siren, 106
  loudness of, 20, 21, 101, 104
  mechanical siren, 109
  power consumption of, 21-22
  resonating horn, 103, 112

Magnetic switch, 5, 28, 29, 30
  installation of, 157-158
Mat switch (see Step mat)
Maximum permissible resistance in perimeter loops, 67

INDEX

McCulloh loop system, 129, 131-135
Metallic foil, 6, 33
   installation of, 153, 156-157
Microphone, 10
Microwave devices, 9

Panic buttons, 20, 34, 70, 81
   circuit for, 70
Pen register systems, (*see* McCulloh loop system)
Perimeter detectors, 5-9, 27-33
   (*see also* Magnetic switch, Step mat, Foil, Vibration switch)
Perimeter loop (supervised), 66
Perimeter switches, 27
Photoelectric principles, 55
Piezoelectric crystal, 51
Police station hook up, 24
   (*see also* Direct line system, Remote alarm systems)
Power consumption, 21-22
Power supply, 18, 83-98 (*see also* Batteries)
Pulse code modulation in wireless alarm systems, 144

Radio (Wireless) systems, 140-144
   amplitude modulation in, 141-142
   F.C.C. regulations for, 140-141
   frequency modulation in, 143
   frequency shift keying in, 143-144
   pulse code modulation in, 144
Recognition equipment for tone coding, 123, 125
Reed relay, 125
Reed magnetic switches, 29
Release signal (in telephone dialer), 116
Remote control stations, 19, 72
Remote alarm systems, 22-25
   locations of, 135-136
   signal, 113
   (*see also* Dialer, Direct line system)
Resonating horn alarms, 103, 112
Reversing relay, 23, 129-130
Ripple, 88-89

Scanners (automatic for closed circuit T.V.), 146
Schmitt trigger, 58
Seismic transducer, 51
   (*see also* Geophones)
Shift register, 60
   (*see also* Flip flop circuit)
Shunt lock, 72
Signal processing circuits, 53, 58
Signal processor, for closed circuit T.V. systems, 147
Silicon controlled rectifiers (S.C.R.), 68-69
Sirens, 18, 20, 99, 106-112
   compressed air type, 109
   electronic, 106-109
   mechanical, 109
   power consumption of, 21-22
Sonic devices, 9, 163
Space intrusion detectors, 5, 9-15, 35-63
   amplifiers for, 48-50
   audio type, 52-53
   differential techniques in, 53-55
   Doppler principle in, 35-41
   false triggering in, 47, 57-58
   F.C.C. regulations for, 39, 41-45
   Gunn diode, use of in, 40
   infrared beams, use of in, 56-57
   installation of, 160-164
   photoelectric type, 13, 55-57
   radiation patterns of, 45-47
   signal processing circuits for, 58-63
   transducers, used in, 50-51
   ultrasonic type, 47-48
Spikes (on power supply lines), 104
Step mat, 32
   mat switch, 6
   installation of, 158-159
Stepping relay, 73
Stress detectors, 12
Stress sensitive transducer, 50
Supervised perimeter loop, 66
   wiring for, 80-81
Switches,
   bullet, 33
   magnetic, 28-30
   perimeter, 27
   reed magnetic, 29

Tamper loop, 110
Tape telephone dialer (see Dialer)
Telephone dialer (see Dialer)
Telephone line, 23
Tilt switches, 33
Time delay mechanism (electronic), 74
Time delay circuits, 19
Time sampling processor (digital), 61
Tone coding recognition equipment, 123
Tone encoded messages, 123
Tools (for installation), 149-150, 164-165
Transducers,
   stress sensitive, 50
   seismic, 51
Transformer, 84, 86
Transients, 86
Trickle charging (of batteries), 93
Trip wire, 33

Ultrasonic devices, 9, 10, 47
Ultra high frequency devices, 9

Vibration contacts, 30
Vibration switch, 7
   installation of, 159-160
Visible light system, 55
Volumetric intrusion detector (see Space intrusion detector)

Wireless (radio) systems, (see Radio systems)

Zone selection circuitry, 78